T0314986

Design for Excellence in Electronics Manufacturing

Wiley Series in Quality & Reliability Engineering

Dr. Andre Kleyner

Series Editor

The Wiley Series in Quality & Reliability Engineering aims to provide a solid educational foundation for both practitioners and researchers in the Q&R field and to expand the reader's knowledge base to include the latest developments in this field. The series will provide a lasting and positive contribution to the teaching and practice of engineering.

The series coverage will contain, but is not exclusive to,

- Statistical methods
- Physics of failure
- Reliability modeling
- Functional safety
- Six-sigma methods
- Lead-free electronics
- Warranty analysis/management
- Risk and safety analysis

Wiley Series in Quality & Reliability Engineering

Design for Excellence in Electronics Manufacturing
Cheryl Tulkoff, Greg Caswell
April 2021

Reliability Culture: How Leaders can Create Organizations that Create Reliable Products
Adam P. Bahret
February 2021

Design for Maintainability
Louis J. Gullo, Jack Dixon
February 2021

Lead-free Soldering Process Development and Reliability
Jasbir Bath (Editor)
September 2020

Automotive System Safety: Critical Considerations for Engineering and Effective Management
Joseph D. Miller
February 2020

Prognostics and Health Management: A Practical Approach to Improving System Reliability Using Condition-Based Data
Douglas Goodman, James P. Hofmeister, Ferenc Szidarovszky
April 2019

Improving Product Reliability and Software Quality: Strategies, Tools, Process and Implementation, 2nd Edition
Mark A. Levin, Ted T. Kalal, Jonathan Rodin
April 2019

Practical Applications of Bayesian Reliability
Yan Liu, Athula I. Abeyratne
April 2019

Dynamic System Reliability: Modeling and Analysis of Dynamic and Dependent Behaviors
Liudong Xing, Gregory Levitin, Chaonan Wang
March 2019

Reliability Engineering and Services
Tongdan Jin
March 2019

Design for Safety
Louis J. Gullo, Jack Dixon
February 2018

Thermodynamic Degradation Science: Physics of Failure, Accelerated Testing, Fatigue and Reliability
Alec Feinberg
October 2016

Next Generation HALT and HASS: Robust Design of Electronics and Systems
Kirk A. Gray, John J. Paschkewitz
May 2016

Reliability and Risk Models: Setting Reliability Requirements, 2nd Edition
Michael Todinov
November 2015

Applied Reliability Engineering and Risk Analysis: Probabilistic Models and Statistical Inference
Ilia B. Frenkel, Alex Karagrigoriou, Anatoly Lisnianski, Andre V. Kleyner
September 2013

Design for Reliability
Dev G. Raheja (Editor), Louis J. Gullo (Editor)
July 2012

Effective FMEAs: Achieving Safe, Reliable, and Economical Products and Processes Using Failure Modes and Effects Analysis
Carl Carlson
April 2012

Failure Analysis: A Practical Guide for Manufacturers of Electronic Components and Systems
Marius Bazu, Titu Bajenescu
April 2011

Reliability Technology: Principles and Practice of Failure Prevention in Electronic Systems
Norman Pascoe
April 2011

Improving Product Reliability: Strategies and Implementation
Mark A. Levin, Ted T. Kalal
March 2003

Test Engineering: A Concise Guide to Cost-Effective Design, Development and Manufacture
Patrick O'Connor
April 2001

Integrated Circuit Failure Analysis: A Guide to Preparation Techniques
Friedrich Beck
January 1998

Measurement and Calibration Requirements for Quality Assurance to ISO 9000
Alan S. Morris
October 1997

Electronic Component Reliability: Fundamentals, Modelling, Evaluation, and Assurance
Finn Jensen
1995

Design for Excellence in Electronics Manufacturing

Cheryl Tulkoff and Greg Caswell

This edition first published 2021
© 2021 John Wiley & Sons Ltd

All rights reserved. No part of this publication may be reproduced, stored in a retrieval system, or transmitted, in any form or by any means, electronic, mechanical, photocopying, recording or otherwise, except as permitted by law. Advice on how to obtain permission to reuse material from this title is available at http://www.wiley.com/go/permissions.

The right of Cheryl Tulkoff and Greg Caswell to be identified as the authors of this work has been asserted in accordance with law.

Registered Offices
John Wiley & Sons, Inc., 111 River Street, Hoboken, NJ 07030, USA
John Wiley & Sons Ltd, The Atrium, Southern Gate, Chichester, West Sussex, PO19 8SQ, UK

Editorial Office
The Atrium, Southern Gate, Chichester, West Sussex, PO19 8SQ, UK

For details of our global editorial offices, customer services, and more information about Wiley products visit us at www.wiley.com.

Wiley also publishes its books in a variety of electronic formats and by print-on-demand. Some content that appears in standard print versions of this book may not be available in other formats.

Limit of Liability/Disclaimer of Warranty
While the publisher and authors have used their best efforts in preparing this work, they make no representations or warranties with respect to the accuracy or completeness of the contents of this work and specifically disclaim all warranties, including without limitation any implied warranties of merchantability or fitness for a particular purpose. No warranty may be created or extended by sales representatives, written sales materials or promotional statements for this work. The fact that an organization, website, or product is referred to in this work as a citation and/or potential source of further information does not mean that the publisher and authors endorse the information or services the organization, website, or product may provide or recommendations it may make. This work is sold with the understanding that the publisher is not engaged in rendering professional services. The advice and strategies contained herein may not be suitable for your situation. You should consult with a specialist where appropriate. Further, readers should be aware that websites listed in this work may have changed or disappeared between when this work was written and when it is read. Neither the publisher nor authors shall be liable for any loss of profit or any other commercial damages, including but not limited to special, incidental, consequential, or other damages.

Library of Congress Cataloging-in-Publication Data

Names: Tulkoff, Cheryl, author. | Caswell, Greg, author.
Title: Design for excellence in electronics manufacturing / Cheryl Tulkoff, US, Greg Caswell, US.
Description: Hoboken, NJ : Wiley, 2021. | Series: Quality and reliability engineering series | Includes bibliographical references and index.
Identifiers: LCCN 2020027863 (print) | LCCN 2020027864 (ebook) | ISBN 9781119109372 (cloth) | ISBN 9781119109389 (adobe pdf) | ISBN 9781119109396 (epub)
Subjects: LCSH: Electronic apparatus and appliances–Design and construction.
Classification: LCC TK7870 .T845 2021 (print) | LCC TK7870 (ebook) | DDC 621.381–dc23
LC record available at https://lccn.loc.gov/2020027863
LC ebook record available at https://lccn.loc.gov/2020027864

Cover Design: Wiley
Cover Image: © graphicINmotion/Shutterstock

Set in 9.5/12.5pt STIXTwoText by SPi Global, Chennai, India
Printed and bound by CPI Group (UK) Ltd, Croydon, CR0 4YY

C9781119109372_160321

I would like to dedicate this book to my wife, June, who was there for me during the entire book-writing process. Thanks my love.

G.C.

I'd like to dedicate this book to my husband, Mike, and my son, David, for their patience and understanding through the long, cranky hours spent writing and rewriting.

C.A.T.

Contents

Contributors

Dr. Craig Hillman
Formerly of Ansys-DfR Solutions

Dr. Nathan Blattau
Ansys-DfR Solutions

Jim McLeish
Formerly of Ansys-DfR Solutions

Randy Schueller
Formerly of Ansys-DfR Solutions

Seth Binfield
Ansys-DfR Solutions

List of Figures

List of Tables

Series Editor's Foreword by Dr. Andre Kleyner

The Wiley Series in Quality & Reliability Engineering was launched 25 years ago. Since then, it has grown into a valuable source of theoretical and practical knowledge in the field of quality and reliability engineering, continuously evolving and expanding to include the latest developments in these disciplines.

Each year, engineering systems are becoming more complex, with new functions and capabilities and longer expected service lives; however, the reliability requirements remain the same or become more stringent due to the increasing expectations of product end users. With the rapid development of autonomous vehicles and growing attention to functional safety, these expectations have grown even further. It will require the utmost reliability to convince people to entrust their lives to an "inhuman machine"; only by using new visions, methods, and approaches to develop engineering systems – and electronic systems in particular – will this become a reality.

The book you are about to read was written by experts in the field of electronics design and manufacturing. Cheryl Tulkoff and Greg Caswell, whom I have the privilege to know personally, have a depth and variety of experiences covering virtually every aspect of design for reliability and quality manufacturing of electronics. This book presents an easy-to-read, step-by-step guide to designing, testing, validating, and building highly reliable electronic systems. It also addresses sustainability and obsolescence – the flip side of fast IC evolution and miniaturization – which are significant issues for electronic systems designed to operate for long periods, such as those in fields such as automotive, airspace, defense, etc.

Despite its obvious importance, quality and reliability education is paradoxically lacking in today's engineering curriculum. Few engineering schools offer degree programs or even a sufficient variety of courses in quality and reliability methods. Therefore, most quality and reliability practitioners receive their professional training from colleagues, engineering seminars, publications, and technical books. The lack of formal education opportunities in this field highlights the importance of technical publications such as this one for professional development.

We are confident that this book, as well as the entire series, will continue Wiley's tradition of excellence in technical publishing and provide a lasting and positive contribution to the teaching and practice of engineering.

Foreword

First and foremost, this book is an invaluable technical gem – front-line experience pours out of it to those (like me) who are listening. This is a work as spoken, not as written or read, from a voice who has seen and knows electronics design, manufacturing, and reliability from the beginning looking forward and the end looking back; and who has gained knowledge from experience, intelligence, and maybe even a few mistakes. Those are laymen's terms for the complex and critical technical and commonsense details that distinguish Designing for Excellence in Manufacturing from the more common "Design it and throw it over the wall to manufacturing" approaches.

Ken Symonds
Technologist
Western Digital Corporation

Preface

With our combined 80+ years of experience in electronics manufacturing encompassing integrated circuit fabrication, printed circuit board fabrication, circuit board assembly, and work in materials, packaging, processes, and standards, we feel uniquely qualified and privileged to author this book.

Our inspiration to write it came from peers, clients, and students in the electronics industry. Through our work with them, we realized we were answering the same questions, solving the same problems, and seeing the industry make the same mistakes over and over. We found that both unfortunate and discouraging. We were unable to find a comprehensive how-to guide and felt we had a unique opportunity to help prevent problems from recurring.

We wrote this book for electronics design, manufacturing, and reliability engineers and those who work with them. We wanted to create a practical guide containing tools, tips, and solutions to help engineers create high-quality, high-reliability products. We sincerely hope we have succeeded.

April 2021
Austin, Texas, USA

Cheryl Tulkoff and Greg Caswell

Acknowledgments

The authors would like to thank the engineering teams at Ansys-DfR Solutions and National Instruments for their support during the development of this book. We would particularly like to thank Dr. Craig Hillman, Dr. Nathan Blattau, Jim McLeish, Dr. Randy Schueller, and Seth Binfield for exceptional technical input in their areas of expertise. Their contributions made the book a more comprehensive resource.

We would like to recognize Ken Symonds for performing a detailed review of the entire book and offering numerous editorial comments that significantly enhanced it. His input on grammar and complexity also made the book infinitely more readable.

Andre Kleyner provided the critical initial nudge and persistent support in driving us to make the book a reality. Ella Mitchell at Wiley secured the technical reviews for us at the concept stage and made sure all of the important paperwork got done.

We'd like to express our gratitude to our peers and students from the electronics industry organizations that we have taught and volunteered for over the years, including IPC, SMTA, iMAPS, IEEE, and ASQ. Their questions, feedback, and knowledge have greatly informed this book.

Finally, we would like to thank our respective companies and families for giving us the time to bring the entire manuscript together. We hope you find it useful.

Cheryl Tulkoff and Greg Caswell

Acronyms

2D	two-dimensional
3D	three-dimensional
AABUS	as agreed between user and supplier
AC	alternating current
ADA	automated design analysis
AF	acceleration factor
AFSC	Air Force Systems Command
Ag	silver
ALT	accelerated life test
AMR	absolute maximum rating
ANSI	American National Standards Institute
AOI	automated optical inspection
ASTM	American Society for Testing and Material
AVL	approved vendor list
BGA	ball grid array
BIST	built in self-test
BOM	bill of materials
BTC	bottom termination component
C	centigrade
C0G	temperature-stable dielectric
CAD	computer-aided design
CAF	conductive anodic filament
CAM	computer-aided manufacturing
CAPEX	component cost and installation expense
CAR	corrective action request
CBGA	ceramic ball grid array

CCA	circuit card assembly
CDM	charged device model
CFF	conductive filament formation
CI	continuous improvement
cm	centimeter
CM	contract manufacturer
CMOS	complementary metal oxide semiconductor
COB	chip on board
COP	computer operating properly
COTS	commercial off the shelf
CPU	central processing unit
C-SAM	C-mode scanning acoustic microscopy
CSP	chip-scale package
CTE	coefficient of thermal expansion
CTQ	critical to quality
CTS	compatibility test suite
Cu	copper
CVD	chemical vapor deposition
DC	direct current
DfM	Design for Manufacturability
DFMEA	design failure modes and effects analysis
DfR	Design for Reliability
DfS	Design for Sustainability
DfT	Design for Test
DfX	Design for Excellence
DI	deionized
DIC	digital image correlation
DIP	dual inline package
DMA	dimethylacetamide; dynamic mechanical analysis
DMF	dimethylformamide
DMSMS	diminishing manufacturing sources and material shortages
DoD	US Department of Defense
DPMO	defects per million opportunities
DRAM	dynamic random access memory
DRB	data-retention bake
DRBFM	design review by failure mode
DRC	design rule check

DSC	differential scanning calorimetry
DUT	device under test
Ea	activation energy
EA	engineering authority
EBIC	electron beam induced current
ECM	electrochemical migration
ECN	engineering change notice
EDA	electronic design automation
EDR	electronics design reliability
EDX	energy dispersive X-ray spectroscopy
EEE	electrical, electronic, electromechanical
EFAT	extended factory acceptance test
EM	electromigration
EMC	electromagnetic compatibility
EMI	electromagnetic interference
EMS	electronic manufacturing services
END	endurance test
ENEPIG	electroless nickel / electroless palladium / immersion gold
ENIG	electroless nickel / immersion gold
EOL	end of life
EOS	electrical overstress
EPDM	ethylene propylene diene monomer
ERC	electrical rule check
ESD	electrostatic discharge
ESR	equivalent series resistance
ESS	electrical stress screening; environmental stress screening
eV	electron volts
F3	form, fit, and function
FA	failure analysis
FAR	Federal Acquisition Regulations
FAT	factory acceptance test
FEA	finite element analysis
FET	field effect transistor
FIB	focused ion beam
FIFO	first in, first out
FIT	failures in time

FMEA	failure modes and effects analysis
FMECA	failure modes, effects, and criticality analysis
FR4	fire retardant 4
FRACAS	Failure Reporting, Analysis, and Corrective Action System
FTA	fault-tree analysis
FTIR	Fourier transform infrared spectroscopy
GBL	y-butyrolactone
GEIA	Government Electronics and Information Technology Association
GHz	gigahertz
GND	ground
Grms	gravity root mean square
HALT	highly accelerated life testing
HASA	highly accelerated stress audit
HASL	hot air solder leveling
HASS	highly accelerated stress screening;
HAST	highly accelerated stress test; highly accelerated stress testing
HATS	highly accelerated thermal shock
HBM	human body model
HCF	high cycle fatigue
HCI	hot carrier injection
HDD	hard disk drive
HDI	high-density interconnect
HPLC	high-performance liquid chromatography
HTOL	high-temperature operating life
I/O	input/output
IC	integrated circuit
ICT	in-circuit test; ion chromatography testing
ID	identification
IDM	integrated device manufacturer
IEC	International Electrotechnical Commission
IEEE	Institute of Electrical and Electronics Engineers
ImAg	immersion silver
IMC	intermetallic growth
ImSn	immersion tin
IPC	association connecting electronics industries

IR	infrared
ISO	International Organization for Standardization
IST	integrated system test
JEDEC	Joint Electronic Devices Engineering Council
JPL	Jet Propulsion Laboratory
JTAG	Joint Task Action Group
KCl	potassium chloride
KPI	key performance indicator
LCCC	leadless ceramic chip carrier
LCD	liquid crystal display
LCF	low cycle fatigue
LCR	inductance-capacitance-resistance
LED	light-emitting diode
LF	lead-free
LGA	land-grid array
LPI	liquid photo imageable
LTS	long-term storage
LU	latch up
MBB	moisture barrier bag
MCM	multi-chip modules
MCM-L	multi-chip module - laminate
MEMS	micro-electro-mechanical system
MFG	mixed flowing gas
MHz	megahertz
MIL	military
MIL-HDBK	military handbook
MIL-SPEC	military specification
MLCC	multi-layer ceramic capacitor
mm	millimeter
MM	machine model
MnO_2	manganese dioxide
MRP/ERP	materials requirements planning/enterprise resource planning
MSL	moisture sensitivity level
MSP	managed supply program
MTBF	mean time between failures
MTTF	mean time to failure
NaCl	sodium chloride

NASA	National Aeronautics and Space Administration
NAVSEA	Naval Sea Systems Command
NBTI	negative bias temperature instability
NDE	non-destructive evaluations
NIST	National Institute of Standards
nm	nanometer
NPI	new product introduction
NRE	non-recurring expense
NSMD	non-soldermask defined
NTF	no trouble found
OBIC	optical beam induced current
ODM	original design manufacturer
OEM	original equipment manufacturer
OPEX	operation/maintenance/intervention expense
ORT	ongoing reliability test; ongoing reliability testing
OSP	organic solderability preservative
Pb	lead
PC	personal computer
PCB	printed circuit board
PCBA	printed circuit board assembly
PCMCIA	Personal Computer Memory Card International Association
PCN	process change notice
PCQR2	printed board capability, quality, and relative reliability
PESD	polymer electro-static discharge device
PFMEA	process failure modes and effects analysis
PGA	pin-grid array
pKa	acid disassociation constant
PLL	phase-locked loop
PoF	physics of failure
PoP	package on package
ppb	parts per billion
ppm	parts per million
PSD	power spectral density
PTH	plated through-hole
PWB	printed wiring board
PWR	power

QBR	quarterly business review
QCI	qualification conformance inspection
QFD	quality functional deployment (house of quality)
QFN	quad flat-pack no-leads
QFP	quad flat-pack
RADC	Rome Air Development Center
RCA	Radio Corporation of America
RDT	reliability demonstration testing
REACH	Registration, Evaluation, Authorization and Restriction of Chemical Substances
RF	radio frequency
RGT	reliability growth test
RH	relative humidity
RIAC	Reliability Information Analysis Center
RMA	rosin, mildly activated
RO	rosin only
ROC	recommended operating conditions
RoHS	Restriction of the Use of Certain Hazardous Substances in Electrical and Electronic Equipment
ROI	return on investment
ROL0	rosin low, flux type L0
ROL1	rosin low, flux type L1
ROSE	resistivity of solvent extract
RPA	reliability physics analysis
SAC	tin/silver/copper (Sn/Ag/Cu)
SAC305	tin silver copper 305
SAE	Society of Automotive Engineers
SAM	scanning acoustic microscopy
SCAR	supplier corrective action request
SEM	scanning electron microscope
SEU	single-event upset
SIG	signal
SIL	safety integrity level
SIMS	secondary ion mass spectroscopy
SiO_2	silicon dioxide
SiP	system in package
SIR	surface insulation resistance
SIT	system integration test

SMD	surface-mounted device
SMT	surface-mount technology
SnPb	tin lead
SoC	system on a chip
SOH	state of health
SPC	statistical process control
SPP	steam pressure pot
SQUID	superconducting quantum interfering device microscopy
Ta	ambient temperature
TA	technical authority
TAL	time above liquidus
TAP	test access points
TBD	to be determined
Tc	case temperature
TC	temperature cycle; thermal cycling
Td	decomposition temperature
TDDB	time-dependent dielectric breakdown
TDR	time domain reflectometry
Tg	glass transition temperature
THB	temp, humidity, and bias
TID	total ionizing dose
TIM	thermal interface material
Tj	junction temperature
TMA	thermo-mechanical analysis
TMF	thermo-mechanical failure
TOF	time of flight
TRAP	technical risk assurance process
Ts	heat sink temperature
TSL	temperature sensitivity level
TSOP	thin small-outline package
TTF	time to failure
TV	television
μg	microgram
UHF	ultra-high frequency
UL	Underwriter Laboratories
UPS	uninterruptible power supply
UUT	unit under test

UV	ultraviolet
V	volt
VAC	volts alternating current
VCC	IC voltage pin
VDR	voltage-dependent resistors
Vds	voltage drain to source
VIP	via in pad
W	watt
WEEE	waste electrical and electronic equipment
WOA	weak organic acid
X7R	temperature-stable dielectric

1

Introduction to Design for Excellence

1.1 Design for Excellence (DfX) in Electronics Manufacturing

Design for Excellence (DfX) is based on the premise that getting product design right–the first time–is far less expensive than finding failure later in product development or at the customer. The book will specifically highlight how using the DfX concepts of Design for Reliability, Design for Manufacturability, Design for the Use Environment, and Design for Life-Cycle Management will not only reduce research and development costs but will also decrease time to market and allow companies to issue warranty coverage confidently. Ultimately, Design for Excellence will increase customer satisfaction, market share, and long-term profits. The Design for Excellence material is critical for engineers and managers who wish to learn best practices regarding product design. Practices need to be adjusted for different manufacturing processes, suppliers, use environments, and reliability expectations, and this DfX book will demonstrate how to do just that.

Design for Excellence is a methodology that involves various groups with knowledge of different parts of the product life-cycle advising the design engineering functions during the design phase. It is also the process of assessing issues beyond the base functionality where base functionality is defined as meeting the business and customer expectations of function, cost, and size. Key elements of a DfX program

Design for Excellence in Electronics Manufacturing, First Edition.
Cheryl Tulkoff and Greg Caswell.
© 2021 John Wiley & Sons Ltd. Published 2021 by John Wiley & Sons Ltd.

include designing for reliability, manufacturability, testability, life cycle management, and the environment. DfX efforts require the integration of product design and process planning into a cohesive, interactive activity known as *concurrent engineering*.

The traditional product development process (PDP) has been a series of design-build-test-fix (DBTF) growth events. This is essentially a formalized trial-and-error process that starts with product test and then evolves into continuous improvement activities in response to warranty claims. DfX moves companies from DBTF into the realm of assessing and preventing issues beyond the base functionality before the first physical prototype has been made. DfX has further evolved as an improvement of the silo approach where electrical design, mechanical design, and reliability work (among others) were all performed separately. DfX allows for maximum leverage during the design stage. Approximately 70% of a product's total cost is committed by design exit. Companies that successfully implement DfX hit development costs 82% more frequently, average 66% fewer redesigns, and save significant money in redesign avoidance. Practicing DfX allows companies to focus on preventing problems instead of solving them or redesigning them.

Each chapter in this book describes and illustrates a specific core element of a comprehensive DfX program. The chapters provide best practices and real-world case studies to enable effective implementation.

1.2 Chapter 2: Establishing a Reliability Program

The chapter will educate you on the core elements of a reliability program, common analysis pitfalls, performing and reviewing reliability data analysis.

At the end of this chapter, readers will be better prepared to:

- Understand the basic elements of a successful reliability program
- Understand the principles associated with reliability analysis and management of the factors that affect product reliability
- Understand the probability density function (PDF) and the cumulative distribution function (CDF) used in reliability
- Understand the reliability prediction models available

1.3   Chapter 3: Design for Reliability (DfR)

The process of Design for Reliability (DfR) has achieved a high profile in the electronics industry and is part of an overall DfX program. Numerous organizations now offer DfR training and tools (sections, books, etc.) in response to market demand. However, many of these are too broad and not electronics-focused. They place too much emphasis on techniques like failure modes and effects analysis (FMEA) and fault tree analysis (FTA) and not enough emphasis on answers. FMEA and FTA rarely identify DfR issues because of the limited focus on the failure mechanism. And they incorporate highly accelerated life testing (HALT) and failure analysis when HALT is testing, not DfR. In addition, failure analysis occurs too late. This frustration with test-in reliability, even HALT, has been part of the recent focus on DfR.

As the design for philosophy has expanded and spread through the electronics marketplace and has become identified with best practices, a diluted understanding of DfR has occurred. True DfR requires technical knowledge of electronics packaging, discrete components, printed boards, solder assemblies, and connectors and how these aspects of electronics can fail under environmental stresses.

This chapter is designed for engineers and manufacturing personnel who need to fully comprehend the characteristics of DfR and how it applies to their unique applications.

1.4   Chapter 4: Design for the Use Environment: Reliability Testing and Test Plan Development

Scientific principles are based on the understanding that products fail when environmental stress exceeds the material strength.

At the end of this chapter, you will:

- Understand the basic elements of a successful reliability testing program
- Understand how reliability testing can be used for the process, part and assembly qualification
- Understand failure patterns based on the ensemble of environmental stressors chosen

- Have a basic understanding of the concepts of accelerated aging rates and acceleration factors
- Understand the difference between a stress screen and an accelerated life test
- Understand the basics of stress-screening equipment
- Have a basic understanding of frequency analysis and power spectral density for vibration and mechanical shock testing
- Have a basic understanding of setting stress levels based on the step-stress algorithm to establish the product operational and destruct limits
- Know how to use the testing process to drive improved reliability in the products manufactured at every site
- Know how to drive robust product design with the testing process and push product performance to the fundamental limits of the material and device technology

1.5 Chapter 5: Design for Manufacturability (DfM)

This chapter provides a comprehensive insight into the areas where design plays an important role in the manufacturing process. It addresses the increasingly sophisticated printed circuit board (PCB) fabrication technologies and processes, covering issues such as laminate selection, microvias and through-hole formation, trace width and spacing, and soldermask and finishes for lead-free materials and performance requirements. Challenges include managing the interconnection of both through-hole and surface mount at the bare-board level. The soldering techniques discuss pad design, hole design/annular ring, component location, and component orientation. You will have a unique opportunity to obtain first-hand information on design issues that impact both leaded and lead-free manufacturability.

1.6 Chapter 6: Design for Sustainability

The best design is not just reliable and manufacturable; products must also be designed with life-cycle management in mind. Designing products to be both reliable and supportable is a critical step in the process. It is one that must be addressed if customers or end-users have long-life, high-reliability, and repairable systems or products.

Key topics covered in this chapter include:

- Obsolescence management
- Long-term storage issues
- Counterfeit prevention and detection strategies
- Baseline life-cycle cost (estimated total ownership cost)
- Use environment verification
- Corrosion protection and mitigation
- Supplier auditing and vendor maturity and stability

1.7   Chapter 7: Root Cause Problem-Solving, Failure Analysis, and Continual Improvement Techniques

Root cause analysis (RCA) is a generic term for diligent structured problem-solving. Over the years, various RCA techniques and management methods have been developed. All RCA activities are problem-solving methods that focus on identifying the ultimate underlying reason a failure or problem event occurred. RCA is based on the belief that problems are more effectively solved by correcting or eliminating the root causes, rather than merely addressing the obvious symptoms. The root cause is the trigger point in a causal chain of events, which may be natural or man-made, active or passive, initiating or permitting, obvious or hidden. Efforts to prevent or mitigate the trigger event are expected to prevent the outcome or at least reduce the potential for problem recurrence.

Effective failure analysis is critical to product reliability. Without identifying the root causes of failure, true corrective action cannot be implemented, and the risk of repeat occurrence increases.

This chapter outlines a systematic approach to failure analysis proceeding from non-destructive to destructive methods until all root causes are conclusively identified. The appropriate techniques are discussed and recommended based on the failure information (failure history, failure mode, failure site, and failure mechanism) specific to the problem.

The information-gathering process is the crucial first step in any failure analysis effort. Information can be gained through interviews with all the members of the production team, from suppliers, manufacturers, designers, reliability teams, and managers to end-users.

Topics to be covered include:

- Root cause problem-solving methodology
- Root cause failure analysis methodology and approach
- Failure reporting, analysis, and corrective action system (FRACAS)
- Failure mechanisms
- Continuing education and improvement activities

The authors hope this book helps you to manufacture your products with better reliability and greater customer satisfaction.

2

Establishing a Reliability Program

2.1 Introduction

A comprehensive, well-thought-out reliability program ensures that companies can achieve their quality, reliability, and customer satisfaction targets on time, on schedule, and within budget. *Reliability* is the measure of a product's ability to perform a required function under stated conditions for an expected duration. By definition, reliability is specific to each application – there is no one-size-fits-all definition. So, it can be useful to start with what reliability is not along with some common myths about reliability.

Myths of Reliability

- Myth 1: Don't worry about design, because most problems are caused by defects from suppliers. While many product failures can be traced back to supplier or manufacturing issues, the most severe warranty issues tend to be design related. Design flaws can affect every product at every customer. As a result, design issues are more likely to result in a recall and have a much more significant impact on a company's bottom line.

- Myth 2: The design is intended for more rugged environments; therefore, nothing can be learned from consumer electronics. The stresses experienced during the operation of a computer or mobile phone can far exceed any loads applied to military, avionics, and industrial designs. For example, laptop computers left in the back of a

Design for Excellence in Electronics Manufacturing, First Edition.
Cheryl Tulkoff and Greg Caswell.
© 2021 John Wiley & Sons Ltd. Published 2021 by John Wiley & Sons Ltd.

car can experience temperatures as high as 80 C on a hot summer day. Combine that with component temperatures that can exceed 100 C during operation and products can be exposed to thermal cycles in number and severity that exceed those experienced in commercial and military applications.

- Myth 3: Design verification is the same as product qualification. The purpose of design verification is to understand the margins of a design. This is typically performed on prototype units using small sample sizes (one to three units is common). Tests performed during design verification include highly accelerated life testing (HALT), corner-case testing, UL testing, ship-shock, etc. Once the design is proven to be robust, product qualification can then be performed. The purpose of product qualification is to demonstrate that design and manufacturing processes are sufficiently robust to ensure the desired quality and lifetime. Product qualification should be performed on a pilot production, not prototypes, and should have a sufficiently large sample size (5 to 20 units) to have some confidence in capturing gross manufacturing issues. There are substantial risks in performing product qualification tests on prototypes. Prototypes that pass qualification may not be representative of production units. This increases the risk of qualification testing failing to capture potential field issues. If prototypes fail qualification testing, these failures may be irrelevant, and attempts at root-cause identification may be a misuse of time and resources.
- Myth 4: Highly accelerated life testing (HALT) can be used to demonstrate product reliability. HALT demonstrates product robustness. Only accelerated life testing can demonstrate reliability. What's the difference? *Robustness* is the measure of a product's ability to withstand stress. For example, one inch of steel is more robust than one mil of paper. This measurement is often defaulted to time zero, which can be either immediately after manufacturing or when the product first arrives at the customer. *Reliability* is the measure of a product's ability to perform a required function under stated conditions for an expected duration.
- Myth 5: Reliability is all predictive statistics. Companies that produce some of the most reliable products in the world spend a relatively insignificant percentage of their product development performing predictive statistical assessments. For example, many original equipment

manufacturers (OEMs) in telecommunications, military, avionics, and industrial controls require a mean time between failures (MTBF) number from their suppliers. MTBF, sometimes referred to as *average lifetime*, defines the time over which the probability of failure is 63%. The base process of calculating MTBF involves applying a constant failure rate to each part and summing the parts in the design. While there have been numerous claims over the years of improvement on this number by applying additional failure rates or modifying factors to consider temperature, humidity, printed circuit boards, solder joints, etc., there are several flaws to this approach. The first is misunderstanding what it means. The average engineer often expects a product with an MTBF of 10 years to operate reliably for a minimum of 10 years. In practice, this product will likely fail far before 10 years. Second, the primary approach for increasing MTBF is to reduce parts count. This can be detrimental if the parts removed are critical for certain functions, such as filtering, timing, etc., that won't affect product performance under test, but will influence product reliability in the field.

Unlike many other elements of the design and development process, reliability requires thinking about failure. For example, successful reliability testing requires *failure*, unlike most other forms of testing, where the goal is to pass.

2.2   Best Practices and the Economics of a Reliability Program

Best Practices

There are no universal best practices. Every company must choose the appropriate set of practices and implement a program that optimizes return on investment in reliability activities. Reliability is all about cost-benefit trade-offs. Since reliability activities are not a direct revenue generator, they are strongly driven by cost. By increasing efficiency in reliability activities, companies can achieve a lower risk at the same cost; and addressing reliability during the design phase is the most efficient way to increase the cost-benefit ratio. Industry rules of thumb indicate the following returns on investment (Ireson and Coombs 1989):

- Issue caught during design: 1× cost
- Issue caught during engineering: 10× cost
- Issues caught during production: 100× cost

Additional Economic Drivers
- Use environment and design life
- Manufacturing volume
- Product complexity
- Margin and profit requirements
- Schedule and delivery needs
- Field performance expectations and warranty budget

2.2.1 Best-in-Class Reliability Program Practices

- Establish a reliability goal and use it to determine *reliability budgeting.*
- Quantify the use environment. Use industry standards and guidelines when aspects of the use environment are common. Use actual measures when aspects of the use environment are unique, or there is a strong relationship with the end customer. Don't mistake test specifications for the actual use environment. Clearly define the median and realistic worst-case conditions through close cooperation between marketing, sales, and the reliability team.
- Perform assessments appropriate for the product and end-user. These assessments require an understanding of material-degradation behavior, either by test to failure or by using supplier-provided data. The recommended assessments include:
  - Thermal stress
  - Margin or safety-factor demonstration (stress analysis that includes step stress tests (e.g. HALT) to define design margins)
  - Electrical stress (circuit, component derating, electromagnetic interference [EMI])
  - Mechanical stress (finite element analysis)
  - Applicable product characterization tests (not necessarily verification and validation tests)
  - Life-prediction validation (accelerated life test [ALT])
  - Mechanical loading (vibration, mechanical shock)
  - Contaminant testing

- Perform design review based on failure mode (DRBFM, Toyota methodology). This readily identifies CTQ (critical to quality) parameters and tolerances and allows for the development of comprehensive control plans.
- Perform Design for Manufacturability (DfM) and Design for Reliability (DfR) and involve the actual manufacturers in the DfM process.
- Perform root cause analysis (RCA) on test failures and field returns to initiate a full feedback loop.

Best-in-class companies have a strong understanding of critical components. Component engineering typically starts the process through the qualification of suppliers and their parts. They only allow bill of materials (BOM) development using an approved vendor list (AVL). Most small to mid-size (and even large) companies do not have the resources to assess every part and part supplier. Those who are best in class focus resources on those components critical to the design. Component engineering, often in partnership with design engineers, also perform tasks to ensure the success of critical components. This includes design of experiments, test to failure, modeling, supplier assessments, etc. Typical critical component drivers are:

- Complexity of the component
- Number of components within the circuit
- Past experiences with component
- Sensitivity of the circuit to component performance
- Potential wearout during the desired lifetime
- Industry-wide experiences
- Custom design
- Single supplier source

From the component perspective, reliability assurance requires the identification of *reliability-critical components* and drives the need to develop internal knowledge on margins and wearout behaviors. RCA requires the true identification of drivers for field issues combined with an aggressive feedback loop to reliability and engineering teams and suppliers.

Best in class companies provide strong financial motivation for suppliers to perform well by creating agreements with the supply chain to accept financial incentives and penalties based on field reliability. These practices allow companies to implement aggressive development cycles, proactively respond to change, and optimize field performance.

Establishing a successful, comprehensive reliability program requires planning and commitment. Requisite priorities include:

1. Focus: Reliability must be the goal of the entire organization and must be implemented early in the product development cycle. Separate reliability from regulatory-required verification and validation activities and mindset.
2. Dedicated staffing: Assignment of responsibilities without assigning resources risks failure.
3. Clearly define reliability goals and the use environment: these drive the rest of the reliability program.
4. Identify critical components, especially those at risk of wearout. Initiate test-to-failure and design-ruggedization activities.
5. Implement step-stress testing at both sub-assembly and assembly levels.
6. Perform RCA. Focus on the top three field issues, and repeat; drive to quality assurance as appropriate.

2.3 Elements of a Reliability Program

Elements of a comprehensive reliability program include:

- Organization: Develop a system that rewards teams for effective reliability engineering focus. Reliability must be the goal of the entire organization and must be implemented early in the process apart from verification and validation (V and V).
- Reliability goals: Consider an availability goal as well.
- Reliability resources: Resources need to be assigned to reliability characterization and committed in program staffing. Characterization infrastructure needs building.
- Software reliability: Frequently overlooked but critical to today's products and systems
- Defined use environments: Engage field service to obtain knowledge of relevant stresses (loads, contaminants, electrostatic discharge [ESD], etc.) in use environment, and then exercise these during reliability characterization.

- Thermal analyses.
- Circuit and component stress analyses.
- Derating: Employ systematic derating by computational and/or experimental analysis.
- Critical components identification: Identify and prioritize components and subassemblies for extended reliability analysis and test activity during early development. Use supplier life data where possible; create the expectation of minimum supplier reliability competency.
- Failure mode and effects analysis (FMEA).
- Critical to quality (CTQs) and tolerance identification.
- Comprehensive control plans with suppliers.
- Design for excellence practices.
  - Design for Manufacturability (DfM), Design for Testing (DfT), Design for Reliability (DfR), Design for Excellence (DfE), and Design for Sustainability (DfS).
  - Manufacturing involvement in DfM and DfT.
- Step-stress tests to define design margins (HALT).
- Simulation for end-of-life prediction
- Relevant product qualification tests
- Accelerated life test (ALT) to validate the life-prediction model
- RCA on failures and field returns with a feedback loop (design, measure, analyze, improve, control [DMAIC])

These elements will be explored in more detail in the upcoming sections and chapters.

2.3.1 Reliability Goals

Desired lifetime and product performance metrics must be identified and documented. The desired lifetime may be defined as the warranty period or by the expectations of the customer. Some companies set reliability goals based on survivability, which is often bounded by confidence levels such as 95% reliability with 90% confidence over 15 years. The advantages of using survivability are that it helps set bounds on test time and sample size, and it does not assume a failure rate behavior (decreasing, increasing, steady-state).

Reliability goals and requirements address the product or system itself and include test and assessment requirements and associated tasks and documentation. These are included in appropriate system/subsystem requirements and specifications, test plans, and contract statements.

Examples of reliability goals and metrics:

- Minimum required field service life under defined conditions: 10 years of environmental exposure *and* 150,000 miles (or 1 million operating hours)
- Maximum allowable quality defect rates
  - Field infant mortality quality defect rates at 3-6-9-12 months
  - Dead on arrival (DOA) defect rate
  - Required minimum field reliability at one year or the end of the warranty period
  - Six Sigma first-pass quality yield
  - Non-repairable or failure to replicate (no trouble found, NTF) rates

2.3.2 Defined Use Environments

Meaningful reliability prediction must consider the environment in which the product is used. There are several commonly used approaches to identifying the environment. One approach involves the use of industry/military specifications such as MIL-STD-810, MIL-HDBK-310, SAE J1211, IPC-SM-785, Telcordia GR3108, and IEC 60721-3. The advantages of this approach include the low/no cost of the standards, their comprehensive nature, and acceptance throughout the industry. If key information is missing from a given industry standard, simply consider standards from other industries. Disadvantages include the advanced age of the standards (some are more than 20 years old) and the lack of validation against current design usage. Depending upon the product and environment, the standards both overestimate and underestimate reliability by an unknown margin.

A second approach to identifying the field environment uses actual measurements of similar products in similar environments. This provides the ability to determine both average and realistic worst-case scenarios. All failure-inducing loads can be identified; and all environments, including manufacturing, transportation, storage, and field, can be included.

An example of where to implement key reliability tools and tests across the product design and development phases is shown in Figure 2.1.

Table 2.1 details common quality and reliability challenges in electronics along with recommended test and inspection methods.

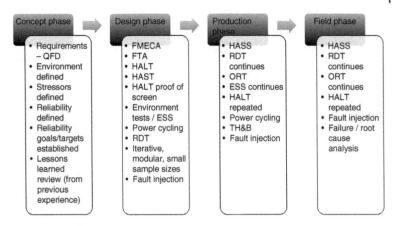

Figure 2.1 Reliability tools across the design and development process.

2.3.3 Software Reliability

The major difference between software and other engineering endeavors is that software is purely design. Hardware unreliability issues are typically related to physical failures of components, which can be traced back to a variety of root causes, such as material, fabrication, or assembly defects; environment or usage overstress; or wearout and design errors. In contrast, when software problems arise, they are traced back to design faults, which come from human errors such as inadequate or incomplete requirements, coding errors, logic errors, improper resource allocation, or the inability of the software to deal with system faults, noise and exception faults.

Therefore, software integrity and dependability are fully dependent on properly implementing quality procedures throughout the specification, design, and coding of the software. This activity should be coupled with meticulous testing and validation activities that can find and correct inadequate requirements, bugs, and discrepancies as early as possible in the design, development, and validation process.

In addition to verifying basic functional operation per the requirements covered in the various industry standards on software development, procedures that also evaluate functional robustness, fault tolerances, and – for devices with operator interfaces – user-friendliness should also be considered. The following procedures are recommended for use

Table 2.1 Common quality and reliability issues.

Area of concern	Impacted item	Failure mechanism	Testing method	Inspection technique
Moisture sensitivity	Plastic IC packages, optocouplers, other polymer-based components	Popcorn delamination at higher reflow temperatures. Heat damage of IC packages	Moisture sensitivity testing J-STD-020	Visual inspection C-SAM
Heat damage	All passive components, circuit boards, connectors	Cracking, dielectric breakdown (capacitors), PCB delamination, warping, or via cracking	Heat resistance, MIL-STD 202G #210F, decomposition temp., time to delamination, package planarity, JESD22-B108	Visual inspection, functional verification
Poor wetting	All solder joints	Cold joints or weak joints fracture in the use environment	Solderability, J-STD-002, J-STD-003	Wetting balance, visual inspection, x-sectioning, lead pull
Solder fatigue	Solder joints, particularly on high-CTE components	Cracked solder joints	Thermal cycling, JESD22-A104, HALT, Vibration	Electrical continuity, visual inspection

Mechanical shock	Solder joints particularly on higher-mass components	Solder joint failure during shipping or handling	Shock test	Electrical continuity, visual inspection
Sn whiskers	Sn and SnCu-plated components	Shorting	iNEMI / JEITA procedures	SEM
Surface mount process control	Insufficient process window creates poor solder joints	Solder joint failures in the use environment	Precondition and assembly, JEDEC 22-A113	X-ray, X-section, visual inspection, reliability test
Rework process	Poor solder joints, damaged components	Joint failures or cracked vias in the use environment	Rework components followed by reliability testing	X-ray, X-section, visual inspection, reliability test
Wave solder process	Incomplete hole fill, fillet lifting, damage to the board	Failed through hole, cracked vias, weak joints	Thermal cycle, HALT Vibration	Electrical continuity, visual inspection
Electrochemical migration	Board surface with no-clean paste residue	Shorting between biased traces in a moist environment	Bellcore GR-78-CORE, J-STD-004 SIR	Visual and resistance after 35 C/85%RH exposure at 50 V

in the specification, creation, development, and validation processes of functional safety of critical software.

2.3.4   General Software Requirements

Requirements Verification Assessment for Accuracy and Completeness
Historically, many software discrepancies can be traced back to specification inadequacies, such as inconsistent, incomplete, imprecise, or vague requirements. Therefore, the first step in any functionality/software validation effort is for the responsible software design organization to perform an analytical review with the specifying organization. The review facilitates an understanding of the requirements, ensures accurate requirements communication, weeds out obvious discrepancies before they are incorporated into the code design, and determines and documents the readiness of the specification for starting the design effort. With this knowledge, corrections can be rapidly implemented.

The review results should be documented in an electronic format (spreadsheet or database) that allows it to be a living document for program tracking, management, and corrective action purposes. It should be organized around the functionality requirement number and title of the software specification. For each requirement line item, there should be a category field or cell to document the following requirement status:

- Requirement is missing.
- Requirement is incomplete or contains to-be-determined items (TBDs).
- Requirement is vague or not understood.
- Requirement is inconsistent or incompatible with other requirements.
- Missing identification of requirements that involve regulatory certification.
- Requirement is not optimized or not user-friendly. Include why and recommendations.
- Requirement is not robust.
- Requirement is inadequate or undefined. Include specifics and recommendations.
- Resolution responsibility and/or owner.
- Date the issue was resolved.

- Resolution comment. Upon resolution, move the original issue(s) to a cleared-issue field, and add a summary of the corrective action.

The knowledge and requirements understanding gained in the specification review should also be used in the creation of the software development validation plan.

Software Development/Validation Plan

The software design organization or supplier is required to develop, maintain, and track performance against a software development validation plan that complies with the requirement of ANSI/IEEE 829, "Standard for Software Test Documentation." Customers may request the use of an alternative, similar format pending approval by the product team. The plan should be in an electronic format (spreadsheet or database) that allows it to be a living document for program tracking and management purposes. The purpose of the software plan is to facilitate an effective test-planning process and to communicate to the entire product, systems engineering, and software team the scope, approach, resources, and schedule of the development validation activities.

The plan also identifies the input/output (I/O), items and features to be tested, testing tasks to be performed, and person responsible for each task. Since it is impossible to test every aspect and I/O configuration of complex software, the plan should also identify any items or features scheduled for limited testing or not to be tested along with the related rationale and risk. Finally, the test plan should also identify and lead to the creation of feature-specific test procedure scripts as needed.

Software Incident Reporting and Tracking

The number of software discrepancy issues for complex programs typically range between several hundred to more than a thousand. To monitor and verify issue resolution and maintain traceability of all incidents, the software design organization should develop a closed-loop database for reporting software incidents and tracking corrective actions that complies with ANSI/IEEE 829 or a similar format.

The database/spreadsheet should be able to generate status reports by sorting and filtering of any category field and content. The database/spreadsheet should generate a monthly histogram bar chart and a monthly mountain chart of issues identification and corrective action performance tracking charts that show the number of:

- New incidents opened
- Total open incidents
- Newly closed incidents
- Total closed incidents

Success Testing Procedures

Success testing (a.k.a. test-to-pass procedures) is the basic preliminary testing designed to verify conformance to requirements. By themselves, these compliance-verification tests do not provide total functionality/software validation because they do not identify the true limits and capabilities of the product or identify many types of faults. (Validation procedures for these type issues are covered in the section "Fault Tolerance and Robustness Testing.") However, just as one would not test a new vehicle at top speed until validation of basic steering, braking, and acceleration functions is achieved, these preliminary tests are vital first steps in the total validation process. The following procedures are recommended for incorporation in the success-testing portion of the software validation plan.

Functionality Requirement Verification via Dynamic Black Box Testing

Dynamic black box testing is an empirical, observational evaluation of inputs and outputs to a device that contains the embedded software or runs a transferable software program. The testing is performed with the device functioning while connected to either a system simulator or a test bench in real time (i.e. dynamic mode). This evaluation is solely based on I/O observation as a result of input stimulation and a combination of I/O states. It is performed without knowledge of how the device works or what the device is doing to process the I/O. It is a basic confirmation of individual CTS functional requirements.

User Interface Testing

User interface testing applies only to the functions of devices that interact with system users/operators. This may be in the form of control panels, displays, alerts, and instrumentation. The objective of user interface testing is to confirm that the user interface functions per CTS requirements. Note that all aspects of user interface validation may not be possible at the supplier. A system simulation bench and/or hardware in the loop testing may be required for some features or aspects. However, it is required that the supplier validate as much of the user interface requirement as possible within the constraints of their systems authority and testing resources.

Software Operational Stability Verification via Dynamic White box Testing

This procedure applies to devices with embedded software. White box testing, sometimes known as X-ray or software structural testing, is a detailed evaluation of how well the software operates internally during actual operation, which may be under either real-time or individual instruction stepping operating conditions. Example of internal software items and functions to be evaluated are:

- Memory management functions and memory integrity
- Interrupt operation and prioritization performance
- Worst-case stack depth penetration
- Power on initialization, shutdown, and reset performance
- Critical code timing performance
- I/O handling and data conversion
- Calibration variables functionality

Software Self-Diagnostics and Trouble Code Function Verification

In addition to the performance evaluations of the device's functional requirements, the self-diagnostic and trouble-code-logging features of the device should also be validated. Discrepancies, faults, and false or overly sensitive diagnostic trouble code triggers are a cause of not-required warranty service with costly "no trouble found" (NTF) warranty events.

Assembly, Service, and Telemetric Interface Feature Verification

All assembly, service, and telemetric interface features should also be validated. Interface faults, initial and service programming and calibration discrepancies and interactive diagnostics are a cause of costly launch issues and NTF warranty events.

Fault Tolerance and Robustness Testing

Fault tolerance testing should demonstrate that the software program can tolerate abnormal or disrupted inputs, power-feed abnormalities, and stressful environment conditions without unanticipated or undesired behavior, such as:

- A fault in one circuit that affects or disrupts any other circuits or the entire system
- A minor disruption that results in the failure of the component to provide useful services
- Permanent crashing of the component
- Lock up or hanging of the component in a busy-loop or anticipation-loop waiting for an expected response or input

Robustness evaluations are in addition to system-level functionality testing that confirms that devices perform their intended function as specified. Robustness evaluations are intended to confirm that the device also doesn't do what it isn't supposed to do, such as shut down or lock up when confronted with a minor abnormality or behave in an unsafe or unstable manner. These procedures are recommended to be performed as development evaluations that accumulate into validation after the majority of functionality performance validation has been performed.

Processor Watchdog Supervisor Performance Evaluation

This procedure is intended to verify that that the systems supervisor circuit was correctly implemented and is effective at recognizing faults and initiating corrective action attempts. Digital microprocessing devices use a "dead man's switch"-like supervision circuit (a.k.a. watchdog or COP (computer operating properly) to monitor for the continued presence of a state of health (SOH) indicator signal. To ensure that disruptions and faults can be rapidly detected and corrected, the supervisor circuit monitors pulses the microprocessor is programmed to send within specified time intervals as the result of handshaking typically between timing interrupt routines and the main programming loop. If the supervisor is not toggled in time, it is assumed that the processor is hung up or executing an endless loop. The supervisor then generates a pulse to the processor to warn that a fault has occurred; typically, this directly or indirectly triggers a system reset that also triggers a diagnostics counter that documents the number of COP-triggered resets over a specified number of system power-up activation cycles.

Fault Injection Testing

Fault injection testing consists of a systematic series of evaluations where hardware and/or software elements are purposefully disrupted or disabled to test and grow the robustness of the whole system to deal with abnormalities and exception faults. The goal is to verify that a device is tolerant of potential system abnormalities. The fault injection procedures focus on functional stability during abnormalities. This requires that:

1. The device will not be physically damaged by an abnormal input or output,
2. The program can recognize fault conditions and abnormal I/O and automatically compensate via alternative operating or graceful degradation operating modes to continue to remain stable and ensure safe system operation to the highest degree possible while issuing fault alerts and logging appropriate diagnostic fault codes.
3. If the abnormality or disruption is removed, the device resumes its normal operating mode.

Before performing this procedure, a mechanization review of the device's internal and external hardware and software is required to organize the device into logical functional subsystems of related input and outputs and to identify the type of fault conditions appropriate to each I/O. This data is to be used to develop a detailed fault injection test script. When function-critical parameters come from digital values, delivered over a data link, the denial or disruption of this data should be included as items in the fault tolerance evaluation plan.

Using Stress Testing

In usage stress testing, the previously defined usage cases should be applied to the device in increasingly faster sequences to evaluate the capability of the software response timing to keep up with highly dynamic usage conditions. The objective of the SW stress test is to determine the robustness of software by testing it well beyond the limits of normal operation. Stress testing is particularly important for mission-critical software but is used for all types of software. Software stress tests emphasize robustness, availability, and error handling under a heavier load than would be considered correct behavior under normal circumstances. Stress testing tries to break the system under test by overwhelming its resources or by taking resources away from it. The primary purpose is to make sure that the software can deal with system overload fails and recover gracefully – this ability is known as *recoverability*.

Worst-Case Testing

Worst-case testing evaluates the software's ability to deal with variation, tolerance stackup and environmental drift in the device's I/O, hardware, and circuits that is short of the total failure situations evaluated during fault injection testing. Worst-case software testing requires a detailed understanding of the design of the device and its software structure. It is the responsibility of the system engineers and software architects to define the specific requirements, test plan scripts, and acceptance criteria for the device. The plan should be reviewed with the product team and the results reported after the testing. The evaluation procedure may reveal that the system requires addition operating state inputs like temperature sensor inputs to identify the extremes of worst-case operating conditions.

2.4   Reliability Data

> Science is an ongoing race between our inventing ways to fool ourselves, and our inventing ways to avoid fooling ourselves.
> *Regina Nuzzo (2015), "How scientists fool themselves –*
> *And how they can stop"*

This quote comes from a fascinating article *Nature Magazine* published in 2015 about how scientists deceive themselves and what they can do to prevent it. *Nature* was responding to a rash of articles decrying bias, reproducibility, and inaccuracy in published journal studies. Although the original focus of the articles concerned the fields of psychology and medicine, the topic is directly applicable to the electronics field and especially relevant to professionals performing failure analysis and reliability research. Reliability data has always been extremely sensitive both within and between companies. You'll rarely see reliability data unless the results are overwhelmingly positive or resulted from a catastrophic event. Furthermore, the industry focuses more on how to organize and analyze data and less about the best way to select or generate that data in the first place. Can you truly rely on the reliability data you see and generate?

Relevant bias recognition and prevention lessons should be learned and shared. For example, how many times have you been asked to analyze data only to be told the expected conclusion or desired outcome before you start? The term *bias* has many definitions, both inside and outside of scientific research. The definition we prefer is that bias is any deviation of results or inferences from the truth (reality) or the processes that lead to the deviation (Porta 2014). An *Information Week* article sums up the impact of data bias on industry well, stating: "Flawed data analysis leads to faulty conclusions and bad business outcomes" (Morgan 2015). That's something we all want to avoid. Biases and cognitive fallacies include:

- Confirmation bias: A wish to prove a certain hypothesis, assumption, or opinion; intentional or unintentional
- Selection bias: Selecting non-random or non-objective data that doesn't represent the population
- Outlier bias: Ignoring or discarding extreme data values
- Overfitting and underfitting bias: Creating either overly complex or overly simplistic models for data
- Confounding variable bias: Failure to consider other variables that may impact cause and effect relationships
- Non-normality bias: Using statistics that assume a normal distribution for non-normal data

Another particularly useful definition comes from the US government's Generally Accepted Government Auditing Standards. They use the concept of *data reliability*, which is defined as "a state that exists when data is sufficiently complete and error-free to be convincing for its purpose and context." Data reliability refers to the accuracy and completeness of data for a specific intended use, but it doesn't mean that the data is error-free. Errors may be found, but errors are within a tolerable range, assessed for risk, and found to be accurate enough to support the conclusions reached. In this context, reliable data is:

- Complete: Includes all the data elements and records needed
- Accurate: Free from measurement error
- Consistent: Obtained and used in a manner that is clear and can be replicated

- Correct: Reflects the data entered or calculated at the source
- Unaltered: Reflects source and has not been tampered with

So, don't simply ask "Is the data accurate?" Instead, ask "Are we reasonably confident that the data presents a picture that is not significantly different from reality?"

Shedding further light on the topic of bias in scientific data and research are some foundations that have made it their mission to improve data integrity and study repeatability. Two such organizations are the Laura and John Arnold Foundation (LJAF) and the Center for Open Science (COS). The LJAF's Research Integrity Initiative seeks to improve the reliability and validity of scientific research across fields that range from governmental to philanthropy to individual decision making. The challenge is that people believe that if work is published in a journal, it is scientifically sound. That's not always true since scientific journals have a bias toward new, novel, and successful research. How often do you read great articles about failed studies?

LJAF promotes research that is rigorous, transparent, and reproducible. These three tenets apply equally well to reliability studies. Studies should be:

- Rigorous: Randomized and well-controlled with sufficient sample sizes and durations.
- Transparent: Researchers explain what they intend to study, make the elements of the experiment easily accessible, and publish the findings regardless of whether they confirm the hypothesis.
- Reproducible: Repeating the work and validating that the outcome is consistent and can be reproduced independently.

The Center for Open Science also has a mission to increase openness, integrity, and reproducibility of research. COS makes a great analogy to how a second-grade student works in science class: observe, test, show your work, and share. These are also shared, universal values in the electronics industry, but things get in the way of living up to those values. COS advocates spending increased time spent on experiment design. This involves communicating the hypothesis and design, having an appropriate sample size, and using statistics correctly. Taking time to do things right the first time prevents others from being led down the wrong path. COS also emphasizes that just because a study doesn't give the desired outcome or answer doesn't make the study worthless. It doesn't even mean that a study is wrong. It may simply mean that the problem being studied is more complicated than can be summed up in a single experiment or two.

Ultimately, ignoring data and analysis biases can lead to catastrophe. The *Harvard Business Review* published a paper (Tinsley et al. 2011) with case studies illustrating the harmful impacts of bias. The Toyota case study shows the consequences of outlier bias. Ignoring a sharp increase in sudden acceleration complaints, the "near misses," led to tragedy. The Apple iPhone 4 antenna example illustrates asymmetric attention bias. The problem with signal strength was well-known and ignored since it was an old problem that had been tolerated by the public – until it wasn't. So, now that some of the many biases and de-biasing techniques out there have been discussed, is your reliability data reliable? How confident are you that it truly reflects reality (Tulkoff 2017)?

2.4.1 Sources of Reliability Data

Sources of reliability data include:

- Suppliers
- Repair, warranty, field
- Development tests: compliance and reliability tests
- Production tests
- Modeling and prediction
- Customer feedback, surveys
- Failure analysis reports
- Industry reports

2.4.2 Reliability Data from Suppliers

Getting reliability data from suppliers is a great way to gain experience in using and analyzing reliability data. Suppliers are a good source for ideas and practices that are standard for their commercial off-the-shelf (COTS) components. Suppliers are also a free or inexpensive source of education on quality and reliability data analysis. Ask for and use them for this expertise.

What kind of data is expected? The specific tests and quantity of data available depend on the industry. Be aware of the relevant industry standards and their requirements; but, also be aware that these standards are "least common denominators." Simply because a part meets a standard doesn't guarantee it will be reliable in each application. The same part may have very different probabilities of failure under different use conditions. At a minimum, expect data on the component life at defined environmental and design conditions (specifications). What data can be requested? Plenty! Surprisingly few customers ask for data that is not volunteered but is often readily available. Ask for any quality and reliability data available for the parts of interest. When and how frequently should data be reviewed? Reliability data should be reviewed at initial part selection; when a part, process, or product changes; when a problem or failure occurs; and on a routine schedule for parts identified as critical to quality (CTQ) or performance. Who performs the testing and analysis? Analyses may be supplier performed, user performed (acceptance-based testing), and/or performed by a third party or independent laboratory. Users should have CTQ suppliers under a scorecard process and should make reliability and quality data part of the supplier-selection and -monitoring process.

Areas to consider for supplier data and supplier performance monitoring include:

- Statistical process control (SPC) report data
- Outputs from a Continuous Improvement (CI) system
- Abnormal lot control data
- Process change notification (PCN) for all changes
- Process change approval (PCA) on all major changes (custom-designed parts)
- Yield and in-line monitor data
- Facility audit results
- ISO 9001 certification (or other relevant certifications or standards)
- Reliability monitor program data
- Storage and handling data

For silicon die and packaging, examples of reliability data that may be available include data retention bake (DRB), electrostatic discharge (ESD), endurance test (END), HAST, high-temperature operating life test (HTOL), latch up (LU), steam pressure pot (SPP), and temperature cycle (TC).

2.5 Analyzing Reliability Data: Commonly Used Probability and Statistics Concepts in Reliability

Reliability probability and statistics is a complex, diverse area that people spend years mastering, so don't expect to get the analysis right the first time through! Properly applied statistics can help clarify an issue; but if misused or misapplied, they can generate misleading results. So be skeptical and pessimistic regarding reliability data. Reliability statistics require careful, honest interpretation. Statistical probability should be used only when there is a lack of knowledge about a situation and if the knowledge cannot be obtained at a reasonable cost in a reasonable amount of time. In other words, when trying to determine a course of action, the best path is to acquire knowledge. Do not rely primarily on predictive statistics and probabilities. Use these tools only as a last-resort strategy or in conjunction with other tools. Don't gamble with product reliability. Much like the stock market, past performance *does not* guarantee future results. Since excellent resource material on reliability statistics exists (Abernethy 2000; Wunderle and Michel 2006, 2007), this section intends to reintroduce and provide a brief refresher of some key concepts, basic recommendations, and common pitfalls.

Reliability statistics are used to describe samples of populations. If the sample size is small, every member can be tested, and representative statistics are not needed. For reliability statistics to be valid, samples must be chosen randomly, and every member must have an equal chance of being selected. For example, when testing to failure, if testing is ended before all items have failed, the sample is not random. The data is referred to as *censored data*. Censored data may use Type I censoring (time censored) or Type II censoring (failure censored). Censored data must be analyzed using special techniques.

Here are some views on statistics from masters of their fields:

> A statistical relationship, however strong and however suggestive, can never establish a causal connection. Our ideas on causation must come from outside statistics, ultimately from some theory.
>
> ***Kendall and Stuart, The Advanced Theory of Statistics***

Ultimately, all our understanding should be based upon real knowledge (scientific, human, etc.). The statistical methods can provide tools to help us gain this knowledge.

Patrick O'Connor, Practical Reliability Engineering

When you can measure what you are speaking about and express it in numbers, you know something about it.

But when you cannot measure it, when you cannot express it in numbers, your knowledge is of a meager and unsatisfactory kind. It may be the beginning of knowledge, but you have scarcely in your thoughts, advanced it to the state of science, whatever the matter may be.

Lecture to the Institution of Civil Engineers, 3 May 1883, William Thomson – Lord Kelvin

Common statistical pitfalls include sample-size errors, distribution or model errors, and failure-to-validate-data errors. Using sample sizes too small to be statistically significant or valid (and using that data to make long-term decisions!) is a common error in high-reliability systems design due to availability, time, complexity, and cost of hardware and testing. Distribution or model errors are also frequently seen. Using normal distributions for non-normal data is often done due to the ease of calculations and over-reliance on easy spreadsheet statistics. Using limited, scrubbed, or unvalidated data is also common. Repeating analyses and modeling at various intervals are highly recommended as more and more credible data is obtained. For example, consider repeating the analysis after product release, after a specific number of product builds, after a certain volume has been manufactured, and after some time in the field.

2.5.1 Reliability Probability in Electronics

Any event has a probability of occurrence between 0 and 100%, or zero to one. A probability of zero means the event will never occur, while a probability of one means the event is guaranteed to occur. Calculating probabilities requires that the data be unbiased, random, come from a representative sample, and use a valid sample size.

Probability event types may be simple or compound. Simple events cannot be broken down, while compound events are composed of two or more events. Compound events can be additive or multiplicative.

- Additive events are composed of independent events. Example: Two cars starting on a cold day, where each has a probability of starting of 0.90. What is the probability that at least one will start?
  - Probability $= P(1) + P(2) - P(1 \cup 2) = 0.90 + 0.90 - (0.90 \times 0.90) = (1.80 - .89) = .91$
- Multiplicative events are composed of dependent events One IC in a circuit has a probability of working of 0.99; another IC in a circuit has a probability of working of 0.90. What is the probability that the circuit will work?
  - Probability $= P(1) \times P(2) = 0.99 \times 0.90 = 0.89$

2.5.2 Reliability Statistics in Electronics

First, identify what type of characterization is needed for the application. Questions to consider include:

- Is a rate being modeled?
- Is there a specific number of trials?
- Is the probability of success the same for all trials?
- Are there discrete functions requiring analysis: pass/fail, working/non-working, on/off?
- Are there continuous functions requiring analysis: controlled by a continuous variable like time?
- Are there point functions requiring analysis: repairable systems where more than one failure or type may occur over time?

Frequently used reliability statistics for electronics include:

- Failure rate = failure per unit of time for a population. Examples:
  - 4 failures (population) per million operating hours $= 4 \times 10^{-6}$ hours.
  - 1000 items operating for a year before failure = 1000 hours $\times$ 365 days $\times \frac{24 hours}{day} = 8{,}760{,}000$ operating hours.
  - Note: In some texts, the term *failure rate* is reserved for repairable systems (more than one failure per device is possible) and the term *hazard rate* (H) is used for non-repair able systems (only one failure per device is possible). In practice, the terms are used interchangeably, with the term *failure rate* commonly used in the US.

- Mean time between failure (MTBF). Example:
  - MTBF = 1 / failure rate
  - Using the failure rate, the MTBF is $1 \div 4 \times 10^{-6}$ hours = 250,000 hours.
- Percent or probability of survival or failure at a point along a timeline. Examples:
  - Reliability = 99.7% (0.9970) at 1 year, 98.9% at 3 years (warranty), 96.5% at 10 years (design life).
  - Failure = (1 − reliability) or (100 − *reliability*%). So, failure is equal to
    * 100 − 99.7 = 0.3% (.0030) at 1 year.
    * 100 − 98.9 = 1.1% at 3 years (warranty).
    * 100 − 96.5 = 3.5% at 10 years (design life).
  - 3 failed out of 1000, 11 failed out of 1000, 36 failed out of 1000.
- The FIT (failure in time) rate is defined as the expected number of component failures per billion (10^9) hours:
  - The FIT rate can be easily converted to the MTBF in hours where: MTBF = 10^9 hours/FIT.
  - The annualized failure rate (AFR) is calculated as AFR = (FIT × $8760 \frac{hours}{year}$)/10^9 hours.
  - The advantage of using FIT rates rather than MTBF is that FIT rates are additive. For example, if the FIT rate of Part A is 125, and the FIT rate of all other component failures is 75, the FIT rate of the system is 125 + 75 = 200.
  - A FIT rate of 200 corresponds to an MTBF of (10^9 hours / 200) = 5×10^9 hours. (5×10^9 hours is a very low failure rate.)
  - A FIT rate of 200 also corresponds to an AFR of (200 hours × $8760 \frac{hours}{year}$)/10^9 hours = 0.00175 = 0.175%.
- DPPM = Defective parts per million
  - Example:
  - 20 pieces are defective in a lot of 1000 pieces. The DPPM is equal to (20 / 1000) =.02 or 2.0% defect ive. 0.020 × 1,000,000 = 20,000 DPPM.

2.5.2.1   Basic Statistics Assumptions and Caveats

Statistics are used to describe samples of populations. If the sample size is small, every member of the population can be tested, and statistics are not needed. Consider ahead of time whether the tool or technique being applied to a data set is appropriate. Many statistical tools are valid only if the data is random or independent. In most cases, samples must be chosen randomly – every member must have an equal chance of being selected. When testing to failure, if the testing is ended before all items have failed, the sample is not random. In this case, the data is censored. Type I censoring is time-based. Type II censoring is failure-based. Censored data can be analyzed using special techniques. Testing of a small lot of parts is typically not effective for detecting issues occurring at a rate below 5% of the population, and the ability to detect all quality and reliability issues by testing alone is simply not practical. This is especially true when the primary objective is the lowest possible cost.

2.5.2.2 Variation Statistics

Variation statistics that most are familiar with are those for central tendency, which include the mean, median, and mode. Dispersion/spread statistics include range, variance, standard deviation, skewness, and kurtosis. The cumulative distribution function provides the probability that a measured value falls between $-\infty$ to $+\infty$. The reliability function provides the probability that an item will survive for a given interval. The hazard function provides the conditional probability of failure in a range given that there was no failure by a certain value (time). A higher failure rate gives rise to a greater probability of impending failure. This is also called the *instantaneous failure rate*.

2.5.2.3 Statistical Distributions Used in Reliability
Discrete Distributions

The binomial distribution is used when there are only two possible outcomes: success/fail, go/no go, good/bad. Random experiments with success or failure outcomes are often referred to as *Bernoulli trials*. For the analysis to be valid, the trials must be independent, and the probability of success for each trial remains constant. The distribution is very useful in reliability and quality work and can model the probability of getting X good items in a sample of Y items. The Poisson distribution is used to model rates of events that occur at a constant average rate with only one of two outcomes countable. It models the number of failures in a given time and describes the distribution of isolated events in a large population. As an example, if the average number of defective units per year per 10,000 units is 0.05, you can evaluate the probabilities of zero, one, two or more faults occurring in a 5-, 10-, or 15-year period.

Continuous Distributions

Commonly used continuous distributions include normal (Gaussian), log-normal, exponential, gamma, X2, t, F, and extreme value. The normal distribution is most frequently (and often incorrectly) used. It's selected because the math is easiest and it's the default for spreadsheet tools. The normal distribution is typically good for quality control activities like SPC. Reliability applications include calculating the lifetimes of items subject to wearout and the sizes of machined parts. The log-normal distribution is usually a better fit for reliability data; simply transform the data by taking a logarithm. It's also a good distribution to use for wearout items. The exponential distribution is extremely important in reliability work; it applies well to life-cycle statistics and assumes a constant hazard or failure rate. The gamma distribution is used in situations where partial failures can exist, such as when multiple subsystems must fail before the full system fails.

The X2, t, and F sampling distributions are used for statistical testing, fit, and confidence and used to make decisions, not to model physical characteristics. The Weibull distribution is the most widely used distribution in reliability applications, especially at the end of the infant mortality period. Adjusting the distribution shape parameter makes it fit many different life distributions. Finally, the extreme value distribution is used when the concern is with extreme values that can lead to failure. It is less concerned with the bulk of a population.

2.6 Reliability Analysis and Prediction Methods

Collecting and analyzing data on product performance, customer satisfaction, warranty, and field repairs is an essential element of a reliability program. Problems do not go away on their own if ignored. So, how can you know what to fix? How do you know if you are working on the most critical problems? How can you avoid the cost of building products with recurring problems that will have to be repaired later? That's where reliability analysis and prediction methods come in. Methods of reliability analysis and prediction include Pareto analysis, MTBF (BOM-based, parts count), reliability growth modeling (design-build-test-fix), CUSUM (cumulative sum charts) and trend charting, block diagrams, and most recently, automated design analysis (ADA) reliability physics predictions using software. A few of these methods are highlighted next.

Pareto Chart

The Pareto chart is a standby classic. It is a histogram that rank orders categories from smallest to largest values (or vice versa) and provides for rapid visual identification of problem-causing issues.

MTBF

MIL-HDBK-217, "Military Handbook for Reliability Prediction of Electronic Equipment," paved the way for MTBF calculations. The handbook was based on statistical and actuarial research work done by the Reliability Analysis Center at the Rome Laboratory at Griffiths AFB, Rome, NY and became required due to Department of Defense backing. Widely used as a purchase requirement for all military electronic applications, it spawned several civilian versions. It contains "statistical" constant (random) failure rate models for electronic parts like ICs, transistors, diodes, resistors, capacitors, relays, switches, connectors, etc. The failure rate models were based on statistical projections of the "best available" historical field failure rate data for each component type. Many assumptions were used, including the belief that past data can be used to predict future performance. It is also known as the *parts counting technique*. To perform an MTBF calculation, select the failure rate for each part in the design from the historical data and apply a thermal stress factor intended to account for diffusion, a solid-state failure mechanism that plagued early electronics (1960s–1980s). Then, calculate the R(t):

$$R(t) = e^{-\lambda t} \text{ and } Rt = R1 \times R2 \times R3... \times Rn \tag{2.1}$$

MTBF was intended to enable reliability comparison of different design alternatives, foster the use of the most reliable components, drive reliability design improvement and determine expected failure rates for military logistics planning (a key driver). However, MIL-HDBK-217 was eventually proven to be both inaccurate and misleading. It frequently caused design resources to be expended on non-value-added efforts. Key disadvantages of the method included:

- Assumption that past performance predicts future results: technology advancements change assumptions.

- The diffusion failure mechanism was designed out.
- The handbook was not updated to account for new /other failure mechanisms.
- It became obsolete quickly: field data could not be collected, analyzed, and logged rapidly enough.
- Increasing design complexity led to increased costs and time required to perform the analysis.
- Average constant failure rate data did not correlate to actual failures.
- It did not address infant mortality, quality, or wearout issues, where up to 65% of failures occur (O'Connor 2012).

The handbook was declared unsuitable for new designs and phased out in 1994. A University of Maryland study documented its flaws and proposed it be replaced by a reliability physics approach. Avoid the use of MTBF or MTTF as reliability metrics. Manipulating these numbers is easy due to the adjustment of multiple quality factors present in the model. MTBF and MTTF are frequently misinterpreted and provide a better fit for logistics and procurement activities.

Reliability Growth Modeling

The old, traditional Western approach to reliability growth and modeling used a design-build-text-fix (DBTF) cycle. This was essentially a trial-and-error approach where the product was designed, prototypes were tested, failures/defects were discovered, corrections were made in the design, more prototypes were made, etc. Traditional OEMs spent almost 75% of product-development costs on this approach (Allen and Jarman 1999). Shortcomings of this approach include:

- Design issues are often not well defined.
- Early build methods do not match final processes.
- Testing doesn't equal actual customer usage.
- Improving fault detection catches more problems but causes more rework.
- Problems are found too late for effective corrective action; quick fixes are often used.
- Testing more parts and more/longer tests are "seen as the only way" to increase reliability.
- OEMs cannot afford the time or money to test to high reliability.

In DBTF-based product development and validation process, reliability growth continued well into production and the field and was a highly reactive process. Using this approach, design engineers worked independently, then transferred designs either "over the wall" to the next department or external to the company. Eventually, manufacturing had to assemble a product not designed for its processes. Since it was too late to make changes, manufacturing struggled to meet yield, quality, cost, and delivery targets. This required trial-and-error crisis management, followed by launch delays, and then quality and reliability issues. This approach has fallen out of favor and been replaced by a combination of concurrent engineering and reliability physics modeling approaches. The newer approaches have the goal of simultaneously optimizing the design across all the DfX disciplines.

Block Diagrams

Block diagrams are useful for performing system-reliability models and calculations and may be simple, parallel, or redundant. A series reliability system model is used when one failure of one component results in the failure of the system. Let's calculate the reliability of a simple fuel system, Rfs, as shown in Figure 2.2.

- $Rfs(t) = Rfp(t) \times Rfl(t) \times Rfi(t) \times Recu(t) \times Rfw(t)$
- $Rfs(t) = 0.980 \times 0.998 \times 0.985 \times 0.975 \times 0.964 = 0.905 \geq 90.5\%$ reliability
- $F(t) = 1 - Rfs(t) = 1 - .905 = .0945 \geq 9.45\%$ failure

For every 100,000 vehicles built, be prepared for $100,000 \times 0.0945 = 9453$ fuel system repairs by time (t).

Parallel or redundant reliability system models are used where all the items in a parallel branch must fail in order for the system to fail. They can model backup systems used to maintain system availability of critical functions in case the primary system fails. A repair of the failed unit is still required, but critical functionality is maintained: for example, separate brake channels where the loss of one brake circuit degrades brake

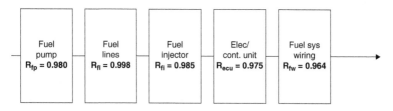

Figure 2.2   Block diagram for a simple fuel system.

Figure 2.3 Parallel brake system.

performance, but reduced braking capability remains. Let's calculate the reliability of a parallel brake system, as shown in Figure 2.3.

- Rbt = 1 − [1 − Rbc1(t)] × [1 − Rbc2(t)]
- = 1 − [1 − 0.990] × [1 −0.990]
- = 1 − [0.010] × [0.010]
- = 1 − 0.0001 = 0.9999 or 99.99%,
- then Fbc = 1 − [1 − 0.990] = 0.02 and Fbt =.0001

For one failure out of two, the calculation is R^2. For two failures out of two, the calculation is $R = 1 − (1 − R)^2$. So, the probability of a single brake circuit failure is $1 − (1 − .01)^2 = .02$, or 2000 incidents for every 100,000 vehicles built. The probability of both brake circuits failing is 0.0001, or 10 incidents for every 100,000 vehicles built: i.e. 200 times less likely to fail due to the dual design.

Automated Design Analysis

Automated design analysis tools represent the latest frontier in reliability analysis and modeling. One example is the ANSYS-DfR Sherlock automated design analysis software. It is a reliability physics-based electronics design software that provides fast and accurate life predictions for electronic hardware at the component, board, and system levels in early design stages. Sherlock allows designers to simulate real-world conditions and accurately model PCBs and assemblies to predict solder fatigue due to thermal, mechanical, and shock and vibration conditions. Approximately 73% of product development costs are spent on the test-fail-fix-repeat cycle. Sherlock design software provides fast and accurate reliability predictions in the earliest design stages tailored to specific materials, components, dies, printed circuit board (PCB)/ball grid array (BGA) stackups, and use conditions. With libraries containing over one million parts, Sherlock reduces FEA modeling time and provides insights before prototyping, eliminating test failures and design flaws, while accelerating product qualification and the introduction of groundbreaking technologies. During preprocessing, Sherlock automatically translates ECAD and MCAE data into 3D finite element models in minutes. In post-processing, Sherlock automates thermal derating and democratizes the thermal and mechanical analysis of electronics.

Sherlock seamlessly integrates with already existing simulation workflows in the hardware design process. It is most valuable when implemented in the early design stages, such as:

- Initial parts selection
- Initial parts placement
- Selecting the final bill of material
- Final layout
- Design for manufacturing

Sherlock makes ANSYS SIwave, ANSYS Icepak, and ANSYS Mechanical users more efficient. It directly connects simulation to material and manufacturing costs.

Additionally, Sherlock's locked IP model protects intellectual property in the supply chain. With the locked IP model, designs are transferred back and forth between design suppliers and design users while preserving PCB design details; the intended use of the PCB design will not be disclosed via environmental conditions or reliability requirements. This communication tool enables two entities to work together on a system with a layer of trust built into the reliability calculations. Sherlock simplifies and improves reliability prediction using a unique, three-phase process consisting of data input, analysis, and reporting and recommendations.

Assessment options include:

- Thermal cycling
- Solder joint fatigue
- PTH fatigue
- Mechanical shock
- Natural frequency
- Harmonic vibration
- Random vibration
- Bending

- Integrated circuit/semiconductor wear-out
- Thermal derating
- Failure rate analysis
- Conductive anodic filament (CAF) qualification
- High-fidelity PCB modeling

Sherlock provides actionable recommendations and presents results in comprehensive, professional reports suitable for internal and external distribution.

2.7   Summary

Establishing a comprehensive reliability program requires extensive planning, training, and expertise. Once established, however, an effective program reduces risk, costs, and failures and results in products that function reliably and satisfy customers. The simple truth is that designing in reliability up front pays off immensely over the life of the product. A reliability program ensures that this happens.

As can be seen, each type of test plan has unique properties and approaches.

References

Abernethy, R.B. (2000). *The New Weibull Handbook* 5e. North Palm Beach, FL: R.B. Abernethy.

Allen, G. and Jarman, R. (1999). *Collaborative R and D.* New York: Wiley and Sons.

Ireson, W. and Coombs, C. (1989). *Handbook of Reliability Engineering and Management.* New York: McGraw Hill.

Morgan, L. (2015). 7 common biases that skew big data results. *Information Week.*

Nuzzo, R. (2015). How scientists fool themselves - and how they can stop. *Nature* 526: 182-185. doi: 10.1038/526182a.

O'Connor, P. (2012). *Practical Reliability Engineering.* New York: Wiley and Sons.

Porta, M. (2014). *A dictionary of epidemiology.* Oxford University Press.

Tinsley, C.H., Dillon, R.L., and Madsen, P.M. (2011). How to avoid catastrophe. Harvard Business Review. https://hbr.org/2011/04/how-to-avoid-catastrophe.

Tulkoff, C. (2017). How reliable is your reliability data? Global SMT and Packaging Magazine.

Wunderle, B. and Michel, B. (2006). Progress in reliability research in the micro and nano region. *Microelectronics and Reliability* 46 (9): 11.

Wunderle, B. and Michel, B. (2007). Printed circuit board design integrity: The key to successful PCB development. http://new.marketwire.com/2.0/rel.jsp?id=730231.

3

Design for Reliability

3.1 Introduction

A long time ago, a wizened professor provided key insights into the beginning of the learning process. In a grand voice, and with a sweeping gesture, he stated, "If you cannot define it, you do not know it." And thus, to honor his legacy, I try to always initiate any educational engagement with a definition of the topic to be discussed. So, to start a chapter on Design for Reliability, let us ask the question: what is Design for Reliability?

Design for Reliability (DfR) is defined as a process for ensuring the reliability of a product or system during the design stage before physical prototype, often as part of an overall Design for Excellence (DfX) strategy.

Each portion of this definition is highly relevant to the topic. A process is a series of actions or steps taken to achieve a specific goal. DfR is not a single activity, but a series of activities that build on each other with the purpose of meeting customer expectations of reliability. Reliability is the measure of a product's ability to:

- … perform the specified function
- … at the customer (with their use environment)
- … over the desired lifetime.

Therefore, the DfR process must take into consideration customer expectations about how and how often the product will operate, how the customer will use the product, and when the customer will no longer blame the manufacturer for the product's failure to operate.

Design for Excellence in Electronics Manufacturing, First Edition.
Cheryl Tulkoff and Greg Caswell.
© 2021 John Wiley & Sons Ltd. Published 2021 by John Wiley & Sons Ltd.

The design stage before physical prototype is self-explanatory. Some will claim that activities after physical prototype, such as highly accelerated life testing (HALT), root-cause analysis, and reliability growth, still provide some value. The counter to that argument is that with all the simulation tools now available, all DfR activities should strive toward optimizing reliability at the design stage. Increasing demand for efficiency is pushing product development from "How many heads per design?" to "How many designs per head?" Within this reality, traditional DfR activities that take up valuable resources, delay market launch, or provide only vague guidance are increasingly out of favor.

Finally, DfR cannot be an island entirely of itself. Reliability, manufacturability, testability, functionality – all of these disciplines significantly influence each other, and a drop-off in one discipline can have a cascading effect on eventual performance and failure to meet reliability goals. DfR should be only one portion of a larger DfX strategy that is in line with the larger goals of the corporation.

So, now that we know what DfR is, why should we do it? What do organizations gain from implementing formal DfR processes? How will DfR help management meet their key performance indicators (KPIs) like time to market, reduced headcount, and lower warranty costs? The benefit of DfR is the same seen with all activities that capture problems before they reach the customer. Management 101 tells us that the earlier problems are captured, the less they cost the organization. There is even a well-accepted industry paradigm that for each stage a problem is missed, the eventual cost is compounded by 10× (Ireson and Coombs 1989). While this guidance typically applies to quality control (both Deming and Toyota were focused on defect reduction), the basic concept holds true for Design for Reliability. Figure 3.1 illustrates this concept.

In addition to these standard cost savings, which apply equally to quality and reliability, DfR provides distinctive additional benefits. The first benefit takes advantage of one of the basic principles of hardware development: obligated costs tend to increase logarithmically, while actual costs increase exponentially. The result is a substantial lag between decisions that affect spending and when that spending hits the accounting ledger.

Speaking of design, the design stage itself can be extremely inefficient. This is especially true for electronics, where electrical designers and mechanical designers compete for functionality, space, and spending.

Figure 3.1 Costs committed vs. money spent.

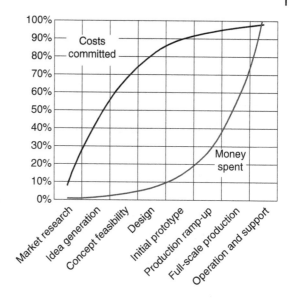

The risk of multiple iterations is so great that many companies in the electronics hardware space plan for these iterations. For example, traditional original equipment manufacturers (OEMs) spend almost 75% of product development costs on test-fail-fix. It makes much more sense to perform a DfR analysis early in the design phase and eliminate the iterations, saving dollars and time. OEMs that use this approach hit development costs 82% more frequently and average 66% fewer re-spins of the circuit board, saving considerable time and money.

3.2 DfR and Physics of Failure

Successful DfR efforts require the integration of product design and process planning into a cohesive, interactive activity known as *concurrent engineering*. Figure 3.2 illustrates this flow.

Initially, it is useful to establish a DfR team consisting of the following members:

- Component engineer
- Physics of failure expert (mechanical/materials)
- Manufacturing engineer

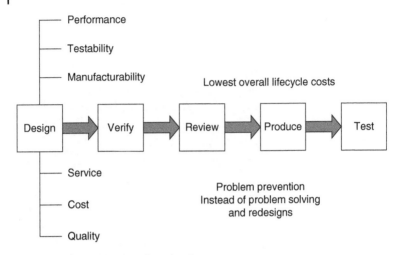

Figure 3.2 Concurrent engineering flow.

- Box level (harness, wiring, board-to-board connections)
- Board/Assembly
- Engineer cognizant of environmental legislation
- Testing engineer (proficient in ICT/JTAG/functional)
- Thermal engineer (depending on power requirements)
- Reliability engineer? This depends. Many classic reliability engineers provide limited value in the design process due to an overemphasis on statistical techniques and environmental testing.

This team performs the functions in Figure 3.3 to develop the product, covering all issues with respect to the design.

This team can then use numerous tools and techniques to assess the reliability of their product. The following list identifies many of these techniques and analyses. (Note that some are specific to industries or applications.)

- Built-in test (BIT) (testability analysis)
- Failure modes and effects analysis (FMEA)
- Reliability hazard analysis
- Reliability block diagram analysis
- Fault tree analysis
- Root cause analysis
- Sneak circuit analysis
- Accelerated testing

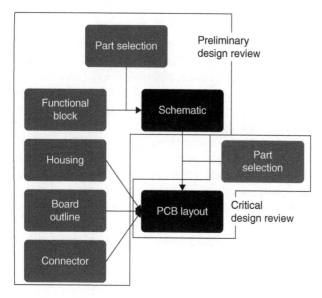

Figure 3.3 Reliability physics in the design phase.

- Reliability growth analysis
- Weibull analysis
- Thermal analysis by finite element analysis (FEA) and or measurement
- Thermal-induced shock and vibration fatigue analysis by FEA or measurement
- Electromagnetic analysis
- Statistical interference
- Avoidance of a single point of failure
- Functional analysis and functional failure analysis
- Predictive and preventative maintenance (reliability centered maintenance [RCM] analysis)
- Testability analysis
- Failure diagnostics analysis (normally part of FMEA)
- Human error analysis
- Operational hazard analysis
- Manual screening
- Integrated logistics support
- Simulation and modeling (stress or damage)
- Derating

Let's look at some of these in more detail.

3.2.1 Failure Modes and Effects Analysis

FMEA is the classic failure mode analysis technique developed after World War II. It forces the team to identify failure modes and their severity, their probability of occurrence, and their detectability and can be executed as both a design analysis (DFMEA) and a process analysis (PFMEA). Conservative, regulated industries love FMEA as they are very concerned about safety and also very concerned about having a written record of being concerned about safety. Other industries are less certain. DFMEA can take too long (a personal computer company completed DFMEA three months after product launch), and PFMEA provided by suppliers can be boilerplate. For an FMEA to be valuable, two things need to happen: (i) the form should be fluid: functional block, geometry, etc. scoring can be linear, actual measurements, etc.; (ii) actions that can be measured through statistical process control should be identified; it is not a one-and-done.

3.2.2 Fault Tree Analysis

This technique is a top-down failure mode analysis approach as opposed to the more typical bottom-up FMEA. It is generally used for root-cause analysis investigations. However, it can be useful in identifying confluence events that could induce an undesirable event. It is also less susceptible to "check the box" syndrome than FMEA.

3.2.3 Sneak Circuit Analysis

This approach is the process of identifying unintended system-level failures that are not due to component failures; that is, the system fails when operating as intended. There are four types of sneak circuits:

- Sneak path: Current/signal flows along an unexpected route.
- Sneak timing: Current/signal flows at an unexpected time.
- Sneak indication: Ambiguous or false display of system operating conditions.
- Sneak label: Ambiguous or false prompt for user action.

3.2.4 DfR at the Concept Stage

Mistakes can be made even as early as the concept stage when considering DfR. Failure to capture and understand product specifications at this stage lays the groundwork for mistakes at schematic and layout. Important specifications to capture at the concept stage include reliability goals, use environment, and dimensional constraints.

As previously noted, reliability is the measure of a product's ability to perform the specified function at the customer (with their use environment) over the desired lifetime. The *desired lifetime* is defined as when the customer will be satisfied with your product. A group of people filed a class-action suit against a television manufacturer, stating that all their TVs failed for the same reason. This turned out not to be the case. However, the important thing is that all the TVs were eight to nine years old – way past the warranty period. But the customers were still upset.

DfR should be actively used in the development of part and product qualification. Methods of assessing reliability might include returns during the warranty period and survivability over lifetime at a set confidence level.

The desired lifetime is also important due to a change in the classic bathtub curve, as shown in Figure 3.4. The new issue involves that of component wearout. For example, integrated circuit (IC) geometries have continued to follow Moore's Law and have continued to shrink, with many manufacturers already producing parts with features that are less than 10 nm. Depending on the environment and life expectancy, these parts may not survive.

Another issue can occur with multi-layer chip capacitor wearout, as shown in Figure 3.5, which depicts capacitor susceptibility. As capacitors

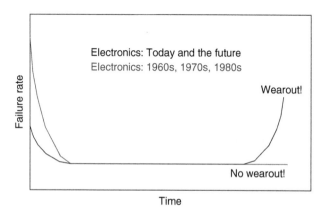

Figure 3.4 Classic bathtub curve.

Figure 3.5 Capacitor susceptibility to wearout and breakdown. Source: Ansys-DfR Solutions.

have continued to increase the capacitance while also lowering the maximum working voltage, the potential for a breakdown between the plates has increased.

Solder wearout can also occur, primarily due to the die-to-package-size ratio. As the ratio has increased, the reduction in cycles to failure solder fatigue has significantly reduced. For example, several years ago, the typical ratio would have been under 40%. However, with the advent of chip scale and other package types, the ratio has moved closer to 75%. The corresponding cycles to failure have gone from over 8,000 to approximately 1,000 due to the change in package stiffness and coefficient of thermal expansion (CTE).

With these issues in mind, the desired lifetime for a new product is a major issue. The following list provides examples of the desired lifetimes for many industries.

- Cell phones: 18 to 36 months
- Laptop computers: 36 to 60 months
- Desktop computers: 36 to 60 months
- Medical (external): 5 to 10 years
- Medical (internal): 7 years
- High-end servers: 7 to 10 years
- Industrial controls: 7 to 15 years
- Appliances: 7 to 15 years
- Automotive: 10 to 15 years (warranty)
- Avionics (civil): 10 to 20 years
- Avionics (military): 10 to 30 years
- Telecommunications: 10 to 30 years
- Solar: 25- to 30-year warranty

Similarly, warranty returns can indicate a product's reliability. Consumer electronics tend to have the highest returns, with desktop PCs being the worst. During the first year, low-volume, high-reliability electronics have a 1 to 2% return rate, while industrial controls have a 500 to 2,000 ppm warranty return rate. This range depends on product complexity, production volumes, and risk sensitivity. Similarly, automotive electronics have a 1 to 5% return rate, while laptop computers have a 1 to 2% warranty return rate that is hardware-related.

Some companies set their reliability goals based on survivability, often bounded by confidence levels (e.g. 95% reliability with a 90% confidence over 15 years.) This approach has both pros and cons. From the advantage perspective, this approach helps to set boundaries on test time and sample size and does not assume a failure rate behavior (decreasing, increasing, or steady-state). A disadvantage, however, it that it can be reinterpreted with mean time to failure (MTTF) or mean time between failures (MTBF) calculations. MTBF and MTTF calculations assume that failures are random in nature, which provides no motivation for failure avoidance. It is also easy to manipulate MTBF numbers: tweaks are often made to reach the desired output (e.g. the quality factors for each component are modified). These actuarial calculations are a better fit for logistics and procurement, not failure avoidance.

The next thing to be understood is the operating conditions (or field environment) that the product will see.

3.3   Specifications (Product and Environment Definitions and Concerns)

The typical first approach for identifying the field environment is to refer to industry/military specifications. There are several to choose from. Some of the available standards include:

- MIL-STD-810
- MIL-HDBK-310
- SAE J1211
- IPC-SM-785
- Telcordia GR3108
- IEC 60721-3

The advantages of this approach are that there are no additional costs. The specifications are often comprehensive; there is agreement throughout the industry; and if information is missing, you can refer to standards from other industries. The disadvantages of this approach are that many of the specifications are more than 20 years old, which makes them not specifically adaptable to current packaging technologies. The results from following this approach don't always lead to accurate results.

A more viable second approach is to fully understand the real-world operating conditions. For this approach, you determine the average and realistic worst-case environments, identify all failure-inducing loads, and address all environments, including manufacturing, transportation, storage, and field.

There are many failure-inducing loads to consider. The following list provides insight:

- Temperature cycling: Tmax, Tmin, dwell, ramp times
- Sustained temperature: T and exposure time
- Humidity: Controlled, condensation

- Corrosion: Salt, corrosive gases (Cl2, etc.)
- Power cycling: Duty cycles, power dissipation
- Electrical loads: Voltage, current, current density, static and transient
- Electrical noise
- Mechanical bending (static and cyclic): Board-level strain
- Random vibration: PSD, exposure time, kurtosis
- Harmonic vibration: G and frequency
- Mechanical shock: G, wave form, number of events

The best practice when addressing this part of a DfR assessment is to use standards when certain aspects of your environment are common or you have no access to the use environment, and to measure when certain aspects of your environment are unique or you have a strong relationship with the customer and can get data. Do not mistake test specifications for the actual use environment. Often, when asked this question, customers will say that their product is good in a −40 C to 85 C operating temperature range, even when it is pointed out that there is no place on earth with that temperature range. Similarly, vibration loads for operation are often incorrect, and the loads of much more stringent tests are used. In both cases, the DfR analysis is impacted by the inaccuracies.

An example of why thinking about all environments is critical can be found in a surgical system for eye care, where the system design incorporated a foot pedal for ease of use. The OEM failed to realize how customers would use the foot pedal. They moved the system across carpet without lifting the foot pedal, which caused large electrostatic charges; and they also used the foot pedal to pull the system, which caused cable/connector failures.

Another example is the way that electronics are exposed in foreign countries. China has issues with grounding in manufacturing facilities, India has issues with numerous brownouts per day and excessive spikes on AC lines, and Mexico has voltage surges. If these environments are not considered during the design phase, then excessive failures are likely.

For electronics used outside with minimal power dissipation, the diurnal (daily) temperature cycle provides the primary degradation-inducing load. Table 3.1 shows diurnal temperature data for Phoenix, Arizona.

Table 3.1 Diurnal temperature for Phoenix, Arizona.

Month	Cycles/ Year	Ramp (hours)	Dwell (hours)	Max temp (C)	Min temp (C)
Jan + Feb + Dec	90	6	6	20	5
March + Nov	60	6	6	25	10
April + Oct	60	6	6	30	15
May + September	60	6	6	35	20
June + July + August	90	6	6	40	25

Also, from an environmental perspective, humidity must be considered, as it can take on several forms:

- Non-condensing: Standard during operation, even in outdoor applications, due to power dissipation.
- Condensing: Can occur in sleep mode or non-powered. Driven by the mounting configuration (attached to something at a lower temperature) and rapid change in the environment. Can lead to standing water if there is condensation on the housing.
- Standing water: Indirect spray, dripping water, submersion, etc., often driven by packaging.

Dimensions are also an issue to be addressed. Keep dimensions loose at this stage. Many hardware mistakes are driven by arbitrary size constraints. Examples include poor interconnect strategies and poor choices in component selection.

A case study of this issue involves the use of 0201 chip components. Tight dimensional requirements pushed the designer toward the wholesale placement of 0201 components, which are not appropriate for technology systems requiring high reliability. Their use caused a major issue with the customer.

Defining the field environment obviously has many elements. Once the ambient conditions are known, the next step is to understand temperature rise within the enclosure. Figure 3.6 illustrates this concept. The influence of ambient temperature on temperature rise can be seen in the figure. At temperatures of 35 C, the maximum temperature rise reaches 30 C. As the ambient temperature drops to 30 C, the maximum temperature rise falls to 20 C. At a lower ambient temperature of

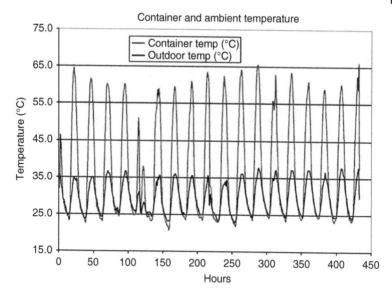

Figure 3.6 Variation in shipping container temperature.

25 C, the maximum temperature rise falls even further, to 10 C. These observations facilitate the correct thermal environment where reliability can be assessed.

3.4 Reliability Physics Analysis

Another tool at the disposal of a reliability engineer in performing a DfR analysis is reliability physics analysis (RPA). This is the process of using modeling and simulation based on the fundamentals of physical science (physics, chemistry, material science, mechanics, etc.) to predict the reliability of a system and prevent failures. Figure 3.7 illustrates how models can be configured to look at the behavior of the circuit board or the vias in the printed circuit board (PCB).

RPA analyses evolved from the term *physics of failure*. The new term is considered more palatable to management and is more descriptive of the practice.

Several things could be modeled, ranging from material changes and movement to defining the magnitude of stresses, the rate the stress is driving the movement, and at what time the material movement will

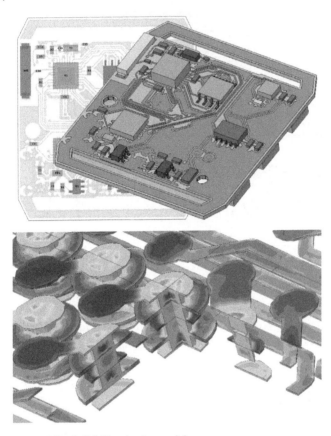

Figure 3.7 Reliability physics models.

induce failure. This process does not use the older MIL-handbooks to predict reliability: those procedures are antiquated, because the data is no longer current. Reliability physics is also not derating criteria or the layout of traces in the circuit board, nor is it the testing performed to validate reliability.

Numerous mechanisms can be assessed through an RPA analysis. They include:

- Electromigration (temperature, current)
- Dielectric breakdown (temperature, voltage)
- Hot carrier injection (voltage)
- Bias temperature instability

- Kirkendall voiding (temperature)
- Galvanic corrosion (temperature, humidity, voltage)
- Low-k dielectric cracking (thermal cycling)
- Wire bond fatigue (thermal cycling)
- Die attach fatigue (thermal cycling)
- Solder joint fatigue (thermal cycling)
- Solder joint fatigue (vibration)
- Solder cracking (mechanical shock, bending)
- Part cracking (mechanical shock, bending)
- Fretting corrosion (temperature, movement)
- Copper migration (temperature)
- Stress relaxation (temperature)
- Oxygen vacancy migration (temperature, voltage)
- Electrolyte evaporation (temperature, current)
- Metallization corrosion (temperature, humidity, voltage)
- Via cracking (temperature)
- Via fatigue (thermal cycling)

RPA should be an integral element of the hardware design process and is shown in Figure 3.8; the shaded areas correspond to where RPA should be performed.

Figure 3.9 illustrates the feedback loop associated with an RPA analysis and how it is integral to defining the life-cycle requirements, impacts materials, and facilitates the system configuration. Module-level suppliers send RPA-based failure model data to system-level customers; system-level customers leverage an RPA model to make data-driven decisions about supplier designs before a physical test and also use the model for system-level optimization.

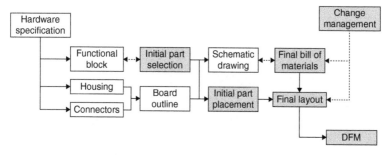

Figure 3.8 Hardware design process.

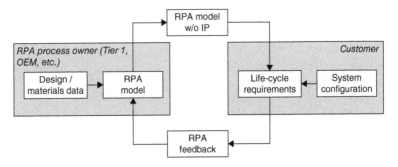

Figure 3.9 Hardware design process feedback loop.

RPA needs to become part of an overall DfR philosophy and process.

Figure 3.10 shows how this is accomplished through deep integration of the process with existing simulation workflows. The figure shows how the electrical, thermal, mechanical, and overall reliability models are integrated to form a cohesive model that supports all the analysis elements.

The defined workflow assumes that a product team controls the design process. In some organizations, research and development create a basic concept or validates new technologies. In this situation, RPA is used even before a hardware specification is generated.

Traditional environmental requirements in hardware specifications are typically insufficient to perform RPA. The focus is on functional performance or limiting the warranty instead of reliability prediction. Environmental specifications should be based on either environmental

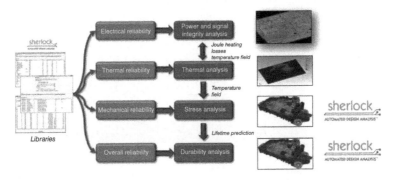

Figure 3.10 Deep integration with existing simulation workflows. Source: Ansys-DfR Solutions.

test requirements or use case(s) that describe realistic worst-case field environments. These could be:

- The main control board shall be able to be stored between -20 C to 85 C.
- The ambient humidity range is 0 to 100% non-condensing.
- The unit shall be operational at an altitude of up to 3,000 meters (10,000 feet). The unit can be transported at an altitude of 12,000 meters (40,000 feet).

Realistic worst-case field environments need to be understood. We have previously defined what the diurnal temperature changes would be for Phoenix, Arizona. Similarly, you would want to define the transportation vibration and shock profiles for the actual application.

RPA also impacts the initial part selection activity, as shown in the following list:

- Which critical components should be subjected to RPA? Within an analog/digital circuit, the critical component is almost always an integrated circuit.
  - Option 1: All critical components, relatively few critical components (5 to 20) in most systems; financially painful if components need to be replaced
  - Option 2: The critical components most likely to fail
- Integrated circuits have three to four different reliability risks: e.g. aging/wearout of silicon transistors (EM, TDDB, HCI, NBTI), cracking of Low-K dielectric, radiation-induced failures of silicon transistors (SEU, TID), solder fatigue of the semiconductor packaging (thermal cycling, vibration). Evaluate critical components based on their susceptibility to these risks.

Now, let's take this a step deeper:

- Aging/wearout of silicon transistors (EM, TDDB, HCI, NBTI): Evaluate all devices fabricated at 90 nm process node or below.
- Cracking of low-K dielectric: Evaluate all devices fabricated at 28 nm process node or below.
- Radiation-induced failures of silicon transistors (SEU, TID): Evaluate all devices fabricated at 28 nm process node or below.
- Solder fatigue of the semiconductor package (thermal cycling, vibration): Evaluate all devices packaged as a ball grid array (BGA), chip-scale package (CSP), quad flat-pack no-lead (QFN), and land grid array (LGA).

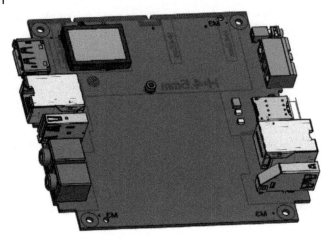

Figure 3.11 Initial parts placement.

The next step is to look at how the part is mounted on the circuit board. Figure 3.11 shows an example of the initial parts placement on the PCB with I/O interaction with the outside world defined (e.g. connectors, switches, buttons, etc.).

The next part of the RPA assessment is to update the thermal analysis through modeling to identify hot-spot locations, as shown in Figure 3.12.

When performing the RPA initial parts placement, it is also necessary to review the keep-out zones; component staking where needed; and any stiffeners required, to assess their benefit for thermal cycling, vibration, and mechanical shock stresses. Doing this at the pre-layout of the PCB reduces design time. This activity can also have a significant impact on

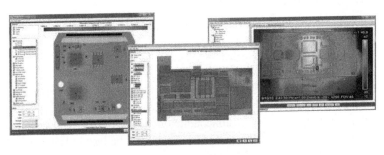

Figure 3.12 Thermal data. Source: Ansys-DfR Solutions.

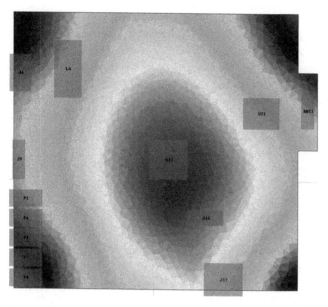

Figure 3.13 Out-of-plane displacement. Source: Ansys-DfR Solutions.

solder joint fatigue. Figure 3.13 shows the displacement impact on a BGA in the center of the PCB. The displacement resulted in a 5% failure rate in 2 years, which was unacceptable. The solution was a combination of moving the part from the center of the board and providing an encapsulant over the part to increase mechanical strength. Knowing this in the design phase saved substantial time and cost.

Next, the RPA analysis focuses on the bill of materials (BOM). The question is, what parts should be subjected to the RPA? They are as follows:

- Any time you have concerns/debates about component derating
- Any discrete components with significant power cycling
- All chip resistors larger than 0805
- All ceramic capacitors rated 6.3 V or below and 105 C or below
- All aluminum liquid electrolytic capacitors

Let's look at an example of a ceramic chip capacitor that needs to perform in Phoenix, Arizona, for 12 years. Earlier in this chapter, Table 3.1 listed the diurnal temperature variation during the year in Phoenix.

An 0402 ceramic capacitor with X5R dielectric has a max rated voltage of 6.3 V, an applied voltage in the application of 5 volts, a capacitance of 10 microfarads, and a rated temperature range of -55 C to 85 C. Will the part survive the 12-year life expectancy for the product? The answer is no. So, the voltage must be lowered, the temperature reduced, or a higher-voltage part chosen to meet the requirement. Knowing this in the design phase saves a significant amount of time and dollars. Similar assessments should be done for electrolytic capacitors and large (2512) chip resistors.

The next operation to be assessed is the circuit board layout, where specific features can be evaluated, such as microvias, the interaction between the circuit board and the housing as a function of thermal cycling.

Finally, the RPA analysis should focus on the Design for Manufacturability (DfM) portion of an overall assessment. This includes things like depanelization, in-circuit test (ICT), heatsink attachment, press-fit connector insertion, daughter card attachment, and placement into the housing. Load limits could be modeled to prevent solder fracture, pad cratering, and component cracking.

3.4.1   Reliability Physics Alternatives

Every organization that designs, makes, or sells electronics has come to realize that reliability is a critical attribute for market success (Hillman 2013). North American, European, and Japanese OEMs repeatedly justify a higher purchase price based primarily on higher reliability (or, at the very least, the claim of higher reliability). Lower-cost newcomers must demonstrate reliability to get beyond the limited percentage of the business-to-business market that cares only about price. In a surprising number of industries, even industries focused on consumer products, a significant share of time and resources goes into reliability assurance activities (design evaluations, simulation/ modeling, prediction, testing, DFMEAs, etc.). By some estimates, over 10% of product costs can be traced directly back to the need to demonstrate reliability. This includes mechanical/electrical/thermal simulations, FMEAs, fault-tree analyses (FTAs), supplier evaluations, test and measurement, and warranty analysis.

But what other avenues of evaluation can be implemented to assist in the reliability physics assessment? The issue is usually how to calculate the MTBF for a system. The very concept of MTBF, and most of our reliability engineering tools, came out of the disastrous performance of military electronics during World War II. By requiring a minimum MTBF number, end users can guarantee the performance of their systems.

There are two big complaints regarding MTBF: how it is determined (prediction) and how it is interpreted (communication). In the electronics industry, MTBF has historically been calculated using empirical prediction handbooks. The mother of all these handbooks is RCA's TR-1100 *Reliability Stress Analysis for Electronic Equipment*, published in 1956. This spawned MIL-HDBK-217, which spawned Telcordia SR-332, IEC-62380, FIDES, and numerous other publications claiming to have the secret to reliability prediction. The failure rates within these handbooks are based on "best available" historical field failure rate data with some "simplifying assumptions."

These handbooks have been shown to have several severe flaws. The original purpose of the original document, MIL-HDBK-217, was not reliability prediction, but to provide a basis for evaluating competing designs. The predictions tend to be overly conservative (one director of engineering described how reliability predictions using MIL-HDBK-217 or SR-332 are divided by 10 or 3, respectively, before being used for planning or marketing purposes). They cannot respond to the pace of technology (parts become obsolete in the time required to gather enough field data to make a prediction). They tend to ignore the mechanical elements of the design (a CPU in a leaded package is viewed as the same as a CPU in an area-array package). And they assume that the failure rate is constant over time, which is not always the case. MTBF is also easily misinterpreted by the outside world. An MTBF of over 50 years can sound impressive until you realize it means almost 20% failure in 10 years (MTBF is not the minimum time to failure). As a result of these drawbacks, there is a big push to eliminate MTBF from the tool set and eliminate it from the vocabulary ("No MTBF"). And, nominally, the authors tend to support such an effort. However, such a movement should be tempered by some of the realities of the world and the needs of complex systems.

There are a number of alternative approaches for risk prediction, including ongoing reliability testing (ORT), warranty analysis, and physics of failure (PoF). However, all these approaches have similar limitations. Most companies do not have the resources to perform ORT. Warranty analysis can sometimes be limited to the first year, when failure rates are still fluctuating. And PoF, as insightful as it is, requires a wearout mechanism. It can't predict an electromagnetic interference / electromagnetic compatibility (EMI/EMC) event, an electrical overstress (EOS) event, or even a mishandling event. For small to mid-size companies with a one-year warranty selling into a data center (a very controlled environment), empirical prediction may be the only option. At the very least, a blended approach may be most effective, where technologies at risk of wearout are modeled using PoF, and other technologies obtain failure rates from other sources, of which one could be a handbook.

While MTBF is sorely misunderstood at the OEM level, it does help managers of complex systems differentiate critical vs. non-critical failures, capture the influence of time to repair, compare to the time during a mission, and perform relatively simple arithmetic to compute availability. As a result, systems people tend to like MTBF and therefore force it on OEMs that supply products that integrate into these complex structures.

Long term, there is a path for the eventual demise of MTBF, especially with more sophisticated reliability tools hitting the market. In the meantime, understanding how and why MTBF is used will bring us closer to that goal. To reach "No MTBF," we need to "Know MTBF."

3.4.2 Reliability Physics Models and Examples

Numerous reliability failure mechanism acceleration models are available to assist you in your DfR analysis based on the failure mode being explored and the materials involved. This section describes several of these (Bayle and Mettas 2010).

3.4.2.1 Arrhenius Equation

The Arrhenius equation is a formula for the temperature dependence of reaction rates. This equation has vast important applications in determining the rate of chemical reactions and calculating energy of activation. Arrhenius provided a physical justification and interpretation for the formula. Currently, it is best seen as an empirical relationship. It can be used to model the temperature variation of diffusion coefficients, the population of crystal vacancies, creep rates, and many other thermally induced processes and reactions. The Eyring equation, developed in 1935, also expresses the relationship between rate and energy.

$$TTF = Ao * e^{(Ea/kT)}$$

where:

- TTF = Time to failure
- Ao = Arbitrary scale factor
- Ea = Activation energy (eV)
- k = Boltzmann's constant
- T = Temperature (Kelvin)

The acceleration factor (AF) is determined by

$$AF = e^{(Ea/k)(1/T_{use}-1/T_{stress})}$$

3.4.2.2 Eyring Equation

$$TTF = Ao * V^{-N*}e^{(Ea/kT)}$$

where:

- TTF = Time to failure
- Ao = Arbitrary scale factor
- V = Voltage
- N = An experimentally determined constant
- Ea = Activation energy (eV)
- k = Boltzmann's constant
- T = Temperature (Kelvin)

The acceleration factor is determined by:

$$AF = (V_{use}/V_{stress})^{-N} * e^{(Ea/k)(1/T_{use}-1/T_{stress})}$$

3.4.2.3 Black's Equation

Black's equation is a mathematical model for the mean time to failure (MTTF) of a semiconductor circuit due to electromigration: a phenomenon of molecular rearrangement (movement) in the solid phase caused by an electromagnetic field. It is most commonly used to develop acceleration factors for photonic devices.

$$TTF = A * J^{-N*(Ea/kT)}$$

where:

- TTF = Time to failure
- A = Prefactor with complex dependence on grain size, line structure and geometry, test conditions, current density, thermal history, etc.
- J = Current density
- N = An empirically determined factor ranging from 1–14. The industry-accepted number today is N = 2.
- Ea = Activation energy (eV)
- k = Boltzmann's constant
- T = Temperature (Kelvin)

The acceleration factor is determined by:

$$AF = (J_{use}/J_{stress})^{-N} * e^{(Ea/k)(1/T_{use}-1/T_{stress})}$$

3.4.2.4 Peck's Law

Peck's Law is an acceleration model for the effect of humidity on the metallized elements of ICs within plastic enclosures (typically an epoxy overmolding). It has been found to be marginal in establishing the acceleration factor for a temperature-humidity-bias test. However, it is still good for getting close.

$$TTF = Ao * RH^{-N} * e^{(Ea/kT)}$$

where:

- TTF = Time to failure
- Ao = Arbitrary scale factor
- RH = Relative humidity (percent)
- N = An experimentally determined constant
- Ea = Activation energy (eV)
- k = Boltzmann's constant
- T = Temperature (Kelvin)

The acceleration factor is determined by:

$$AF = (RH_{use}/RH_{stress})^{-N} * e^{(Ea/k)(1/T_{use}-1/T_{stress})}$$

3.4.2.5 Norris-Landzberg Equation

Norris and Landzberg proposed that the plastic strain range is proportional to the thermal range of the cyclic loading (Delta T). They also modified the equation to account for the effects of thermal cycling frequency (f) and the maximum temperature (T) (Schenkelberg 2018).

This equation is a modification of the older Coffin-Manson equation, which was

$$Nf = Co * (deltaT)^{-N}$$

where:

- Nf: Number of cycles to failure
- Co: A material-dependent constant
- Delta T: Entire temperature cycle range for the device
- n: An experimentally determined constant

The acceleration factor is determined by:

$$AF = (deltaT_{use}/deltaT_{stress})^{-N}$$

Ansys-DfR has done extensive testing to validate the following modifications to the Norris-Landzberg acceleration factor equation to accommodate both Sn/Pb and lead-free (LF) solders:

$$AF = N_{field}/N_{test} = (f - field/f_{test})^n * e^{(Ea/k)(1/T_{maxfield} - 1/T_{maxtest})}$$

where:

- AF is the acceleration factor for the test.
- Nfield is the total number of thermal cycles experienced in the life of the system.
- Ntest is the total number of thermal cycles in the test that will produce damage equivalent to the damage caused by the total number of cycles the system will see in the field.
- ffield is the temperature cycling frequency in the field daily (1 cycle per day for a diurnal temperature cycle).
- ftest is the temperature cycling frequency in the test daily (cycles typically 1 per hour or 24 per day).
- m is the exponent for the temperature cycling frequency ratio (.333 for Sn/Pb and .136 for SAC305).
- delta Ttest is the temperature range experienced in the test.
- delta Tfield is the temperature range experienced in the field.
- n is the exponent for the temperature range ratio (1.9 for Sn/Pb and 2.65 for SAC305.
- Ea is the activation energy (eV) (.121935 eV for SnPb and .188289 eV for SAC305.
- k is Boltzmann's constant.

- Tmax field is the maximum temperature experienced in the field (degrees Kelvin).
- Tmax test is the maximum temperature experienced in the test (degrees Kelvin).

3.4.2.6 Creep Mechanisms

Creep deformations are time-dependent changes under a certain applied load. Thermal creep usually occurs at high temperature but can happen at a slower rate at room temperature. As a result of creep, the metal/material sees a time-dependent change (increase) in length that could be problematic in use.

$$TTF = Bo * (To - T)^{-N} * e^{(Ea/kT)}$$

where:

- TTF: Time to failure
- Bo: Process-dependent constant
- To: Stress free temperature for metal ($\approx$metal deposition temperature for aluminum)
- T: Temperature (degrees Kelvin)
- n: n = 2–3 (n usually 5 if creep)
- Ea: Activation energy (eV)
- k: Boltzmann's constant
- T: Temperature (Kelvin)

The acceleration factor is determined by:

$$AF = (T_o - T_{accel}/(T_o - T_{use})^{-N} * e^{(Ea/k)(1/T_{use}-1/T_{stress})}$$

These equations, and several others, can be used to develop TTF and AF factors for several tests and failure modes.

3.4.3 Component Selection

The next DfR process is that of creating a BOM during the virtual design process (before the physical layout). Figure 3.14 illustrates a typical part/BOM development flow. Many companies also do this during the creation of their approved vendor list (AVL), which is design-independent. Keep it simple. New component technology can be very attractive but is not always appropriate for high-reliability

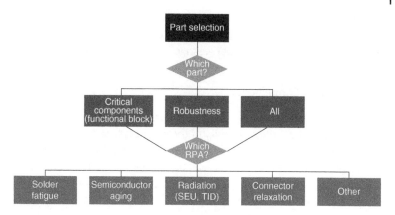

Figure 3.14 Part selection: BOM creation flow.

embedded systems. Be conservative. Also, marketing hype *far* exceeds actual implementation. Component manufacturers often use sales of portable products to boost their numbers. For example, a company claims that "We have built hundreds of millions of these components without a single return." The reality is that the components were sold to two cell phone customers whose product lifetimes were only 18 months.

Even when parts are used by high-reliability companies, some modifications may have to be made. For example, a state-of-the-art crystal oscillator required specialized assembly to avoid failures one to three years later in the field. There are several examples where care should have been taken prior to using a part in a high-reliability application. These include new technologies, X5R dielectric capacitors, silicon carbide diodes, new packaging, quad flat-pack no-lead, 0201, and chip-scale packages.

Another example is that X5R dielectric capacitors with maximum working voltages of 6.3 V or lower are significantly more susceptible to failure due to instability than the same part with either a 10 V maximum operating voltage or X7R dielectric. This resulted in an unintended side effect of the increasing miniaturization of electronic designs: a dramatic reduction in long-term reliability of components that rely on dielectric insulation. While extensive resources have been focused on characterizing semiconductor wearout, the reality is that the current generation of ceramic capacitors can induce failure long before the desired design life.

0201-size components have been found to be more susceptible to dendrites growing across the gap between pads because they are harder to clean under, leading to higher leakage currents and shorting.

3.4.4 Critical Components

Critical components must be addressed during DfR assessment. These include parts where the sensitivity of the circuit is based on component performance and the number of components within the circuit. Often, the output from an FMEA or FTA can have a significant impact on identifying critical components. Industry-wide experience should be monitored, as well as your own experiences. The complexity of a component also defines whether it is critical.

Industry experience has identified optoelectronics as critical parts, as they are manufactured on low-volume assembly lines where high-volume controls are not typically present; it also has been shown that they can wear out in less than 20 years. Similarly, any type of low-volume or custom part can be categorized as critical because it is not a commodity item. Memory devices should also be monitored if the non-volatile memory has limited data-retention time and write cycles. Parts with mechanical movement (e.g. switches, relays, fans, etc.) can wear out quickly depending on the environment. Multi-layer ceramic capacitors (MLCCs) can experience failure as a function of the manufacturing process, where either mechanical or thermal stresses have been known to cause internal cracks in the components.

New technologies must be properly assessed. Often, contract manufacturers have not upgraded their equipment to handle the latest generation of devices: micro-electro-mechanical systems (MEMS), smaller and smaller gate geometries for integrated circuits, green materials, etc. Electronic modules also fit into the critical component category as they are miniature assemblies rather than commodity components. Power components and fuses also are considered critical components.

Electrolytic capacitors have long been identified as a weak link for long-term, high-reliability applications. However, capacitor manufacturers have made significant improvements in materials and manufacturing processes to enhance their reliability. Several environmental factors are involved in the aging of electrolytic capacitors. Electrolyte loss due to drying out and leakage current due to oxide degradation are thermally related, as is self-heating associated with ripple current. The impact of the applied voltage level is also a driver as it can cause leakage-current increases. All these issues result in decreased capacitance, increased equivalent series resistance (ESR), and a change in the dissipation factor. Recently, there have been significant changes in quality-control techniques and the materials used to manufacture electrolytic capacitors. Quality control has improved for foil purity and electrolyte volume. There has also been a significant migration to higher-temperature electrolytes, transitioning from ethylene glycol to dimethylformamide (DMF), dimethylacetamide (DMA), and γ-butyrolactone (GBL). Changes in bung material from butyl to ethylene propylene diene monomer rubber (EPDM) have also occurred. These chemicals have given this rubber compound greater resistance to ultraviolet rays and some chemicals, and it is able to operate in moderate heat.

3.4.5   Moisture-Sensitivity Level

The moisture-sensitivity level (MSL) is detailed in Chapter 5. However, it also needs to be part of the DfR assessment. During the initial design phase, performing an analysis of the different packaging formats in your BOM and determining the MSL level is a good way to eliminate potential manufacturing-related issues later in the process. In general, do not use parts with an MSL rating greater than 3, as it significantly reduces reliability and limits exposure time on the manufacturing floor.

3.4.6   Temperature-Sensitivity Level

Similarly, the temperature-sensitivity levels (TSLs) of the components in the BOM need to be assessed as part of the DfR review. This analysis must look at the maximum operating and storage temperature levels for all components. Many manufacturers have been known to say that components are Restriction of Hazardous Substances (RoHS) compliant and imply that they can handle the higher temperatures associated with LF assembly. This is not always the case, and a TSL analysis as part of the review can identify this issue before a prototype is even manufactured.

3.4.7 Electrostatic Discharge

The risk of electrostatic discharge (ESD) has increased due to the continuous reduction in gate geometries for integrated circuits. From a DfR perspective, ESD is defined as a single-event, rapid transfer of electrostatic charge between two objects, usually resulting when two objects at different potentials come into direct contact with each other. ESD can also occur when a high electrostatic field develops between two objects in proximity. ESD is characterized predominantly by three models: the human body model (HBM), the charged device model (CDM), and the machine model (MM). Recently, additional emphasis has also been placed on interconnection schemes for electronics. The HBM simulates the ESD event when a person charged to either a positive or negative potential touches an IC that is at another potential. The CDM simulates the ESD event wherein a device charges to a certain potential and then comes into contact with a conductive surface at a different potential. The MM simulates the ESD event that occurs when part of a piece of equipment or tool comes into contact with a device at a different potential. The HBM and CDM are more real-world than the MM. For a proper DfR assessment, it is necessary to determine the potential impact of ESD at the IC, component package, and system, as well as the manufacturing process for your electronics.

So, for a DfR analysis, you need to know the ESD rating for each part in your BOM and, if possible, select parts that have the best ESD rating. You need to identify all ESD-sensitive parts on assembly drawings and mark their locations on the surface of the PCB. From a design perspective, place ESD-sensitive components and traces to avoid locations where the board may be handled; consider ESD as well as RF shielding; where possible, install protective devices before ESD-sensitive parts; and avoid coupled ESD events. Do not route traces to ESD-sensitive parts near lines connected to the outside world. Then, perform a circuit analysis to ensure the effectiveness of the protection. If ESD-sensitive parts are used in the design, the circuitry connected to device pins should be evaluated. Ensure that it provides attenuation to prevent voltage in excess of the parts' ESD rating from developing, in case the pin or connected traces are contacted during board handling or system assembly. Figure 3.15 illustrates the various points in a product where ESD could enter.

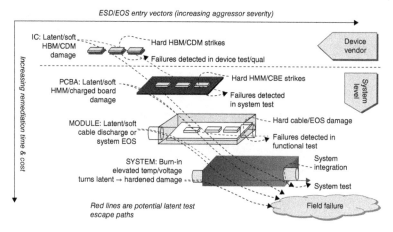

Figure 3.15   ESD entry vectors.

ESD-protection circuitry comes in many forms. Passive networks using capacitors, and band-pass filters provide simple, low-cost protection. For slower-speed devices, Schottky diodes and diode clamping arrays provide good coverage. Low-capacity protection diodes are beneficial for higher-speed circuits; and finally, polymer ESD (PESD) devices provide excellent protection for high-frequency characteristics, have low ignition levels, and can handle multiple ESD pulses without failure.

3.4.8   Lifetime

When ascertaining the viability of components for your design, you must consider any issues associated with the lifetime expectancy of those components. For example, electrolytic capacitors have a history of limited lifetime. To get them to achieve decent lifetimes, you must monitor and derate the voltage at a level 25 to 33% of the maximum and maintain an adequate distance from other components so that their heat does not impact the capacitor. In addition, you must derate the temperature at least 20 C from the maximum rated and design the circuitry to minimize ripple current self-heating. Even with all these actions, the capacitor may not meet your longevity requirements. Typically, the failure mode for electrolytics is the evaporation of the liquid electrolyte through the rubber seal (bung). Evaporation prediction has been based on the standard relationship, which classically has noted that a 10 C drop in temperature will double the capacitor life. This is shown in the following calculation:

$$L_x = L_0 * 2^{(t_0 - t_x)/10}$$

Note that different capacitor manufacturers use slight variations of this calculation. The key difference is the constants used, which tie back to the manufacturer's test data. This often means the data is skewed; you need to take that into consideration when selecting these parts.

A similar calculation can be used for MLCCs. This calculation is as follows:

$$t_1/t_2 = (V_2/V_1)^n expEa/KB(1/T_1 - 1/T_2)$$

where:

- t is time
- V is voltage
- T is temperature (K)
- n is a constant (1.5–7; nominally 4–5)
- Ea is an activation energy (1.3–1.5)
- KB is Boltzmann's constant (8.62×10^{-5} eV/K)

3.5 Surviving the Heat Wave

Since the advent of surface-mount technology back in the 1970s we, as an industry, have continually worked to miniaturize our products. This evolution of product design has impacted us at the semiconductor, package, circuit board, and system levels. So the question is, why do electronics fail under thermal cycling? At the semiconductor level, you can have issues with delamination, IC complexity, degradation mechanisms, associated ceramic capacitor wearout, and EOS. At the package level, issues with bond wires and stacked dies add to the reliability impact. At the PCB level, issues with solder wearout, solder phase coarsening, printed wiring board (PWB) laminates and glass materials, plated through-hole (PTH) fatigue, and the impact of potting can also affect reliability; whereas at the system level, heat sinks and other methods of heat removal can improve the

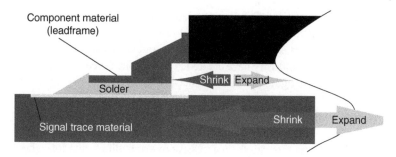

Figure 3.16 Expansion and contraction behavior.

situation. What drives these issues is that we use a variety of materials: e.g. semiconductors, ceramics, metals, and polymers. We then bond them together with other materials like solder and adhesives. Each of these materials has a CTE that is unique and therefore expands and contracts at different rates (Caswell 2017).

Two different expansion/contraction behaviors are shown in Figure 3.16. Because solder is connecting two materials that are expanding/contracting at different rates (global), and because the solder itself is expanding/contracting at a different rate than the materials to which it is bonded (local), we can see how the stresses occur.

The differential expansion and contraction introduce stress into the solder joint that causes the solder to deform (aka *elastic and plastic strain*). The extent of this strain (the *strain range*) tells us the lifetime of the solder joint. The higher the strain, the more the solder is damaged, and the shorter the lifetime. It is then necessary for the person performing the DfR analysis to understand the critical drivers for solder fatigue so that predictive models and design rules can be developed. The parameters that must be understood are the CTE of the components and PWB, the elastic modulus of the components and PWB, the length of the component, the volume and thickness of the solder, and the solder's fatigue properties.

This means the DfR analysis must look at issues from the die level to the system level to be complete. For example, delamination of a die and the complexity of the IC, particularly the gate geometries, can impact wearout and life expectancy. Semiconductor lifetime modeling – where a DfR analysis also looks at the effects of hot carrier injection (HCI), dielectric breakdown, electromigration, and bias temperature instability in the IC – is critical. The degradation mechanisms for each of these characteristics must be understood. Ceramic capacitors are also known to wear out when exposed to high thermal and electrical stresses. So, of course, we as an industry have created higher-capacitance/lower-voltage capacitors with tighter plate tolerances. From a DfR perspective, we usually identify a capacitor as a risk if it has a rated voltage of 6.3 V or lower with X5R or Y5V dielectric. The newer 4 V parts present a similar risk scenario. At the package level, bond wires can present a reliability risk as they fail from excessive intermetallic formation. These intermetallics form at different rates as a function of the bond wire and bond pad metallizations (Breach 2010; Levine and Deley 2004). Similarly, die stacking has created issues with heat trapping (higher junction temperatures). Solder wearout can be impacted by the fact that we have reduced the size of packages with respect to the die inside, which has significantly reduced the cycles to failure. Similarly, we are seeing much hotter devices due to increases in the delta temperature in applications.

At the solder level on the circuit board, we must be concerned with solder coarsening due to thermal stress es. Solder phase coarsening is due to two fundamental forces that drive grains in polycrystalline materials to grow. Grains coarsen because grain boundaries are areas of high potential energy. This is the primary driving force in single-phase materials. During grain coarsening, as grains grow larger, the total grain boundary area decreases, which, in turn lowers the free energy of the system. Smaller particles tend to combine into a larger particle – one with lower total interfacial energy. Grain growth is driven by the local boundary curvature and the presence of triple junctions, which remain in equilibrium and act as anchors to grain boundary mobility. Grain boundary mobility is highly dependent on grain orientations across the boundary. From the original solder joint, grains grow as the solder joint is stressed. The growing grains cause micro-voids to appear on the grain boundaries. These micro-voids connect with each other to create micro-cracks and, eventually, macro-cracks. If the load is distributed evenly across the joint, this happens everywhere at the same time. In BGA balls, this happens in an entire layer of bulk solder. If the load is not even, then the grain growth and micro-cracks are formed in the stress concentration, and the micro-crack advances along a crack path. Figure 3.17 shows the difference between a good cross-section of a solder joint and one where the solder has coarsened and cracked.

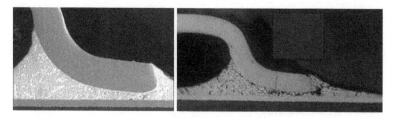

Figure 3.17 Images of solder coarsening. Source: Ansys-DfR Solutions.

PCBs are fabricated with a variety of glass styles that vary the resin content and fiber volume content. The problem is that the datasheets for these materials tend to default to the 7628 glass style, even though complex PWBs typically use a low-volume fraction of glass. A realistic target for the board CTE is between 15 and 17 ppm/degree C. Most laminate suppliers provide CTExy values, while the z-axis expansion, which is only constrained by the PTHs, is more susceptible to expansion issues. This out-of-plane CTE can be 70 ppm/degree C, creating a differential between the copper plating at 17 ppm/degree C. When a PCB experiences thermal cycling, the expansion/contraction in the z-direction is much higher than that in the x-y plane. The glass fibers constrain the board in the x-y plane but not through the thickness. As a result, a great deal of stress can be built up in the copper via barrels, resulting in eventual cracking near the center of the PTH barrel.

Potting can also impact product reliability. One issue we commonly encounter at DfR is that the glass transition temperature (Tg) of the potting is within the minimum and maximum temperature of operation for a product. Near the Tg, the CTE changes more rapidly than the modulus. Changes in the CTE in polymers tend to be driven by changes in the free volume. Changes in modulus tend to be driven by increases in translational/rotational movement of the polymer chains. Increases in CTE tend to initiate before decreases in modulus because lower levels of energy (temperature) are required to increase free volume compared to increases in movement along the polymer chains.

Finally, at the system level, heat sinks can improve the overall situation, as can other methods of thermal relief. Optimal design of a heatsink, meeting program targets for cost, weight, size, and performance, is one of the more challenging activities within most electronics engineering teams.

As you can see, a properly performed DfR analysis can be extensive.

3.6 Redundancy

Redundancy can have a significant impact on system-level reliability, albeit at a cost. A typical system is made up of a variety of component types configured to perform the system application. The failure of a component may lead to the complete failure of the system. This requires, as part of the DfR assessment, an analysis of the component failure modes and mechanisms. In some high-reliability applications, it is necessary to avoid failures that can impact system operation. Often this is accomplished using redundancy. The typical implementation has been to use redundancy by having two separate systems perform the same function (NI 2017).

There are several approaches for implementing redundancy. This section addresses three approaches. Let's first define them.

Standby redundancy is an approach where you have a backup unit that is identical to the original. The second unit acts as a spare until it is needed to take over for the primary due to a failure. Some form of command and control is necessary to effect the transition. The system cost increase will be at least two times the original or more, depending on the software controls. There are two types of standby redundancy. The first is cold standby. In this approach, the secondary unit is turned off, which maintains its reliability. However, the process of bringing the unit online offers some synchronization issues, so there is usually a momentary downtime to deal with. Hot standby is the second approach, where the secondary unit is powered and can be used to monitor the primary unit for issues. In this manner, the transition when failure occurs is much smoother, increasing the availability of the system.

Parallel redundancy is another approach. This method has multiple units running in parallel with all systems synchronized. All systems receive the same data and output the same data. The control system decides which data to use. This approach makes system availability very high.

Dual and *triple* redundancy refer to the number of functional units used. The control decisions about when to switch to the secondary unit are more complex and, as such, drive a cost impact of approximately two times that of the basic system. A triple system has two standby units and drives a three-times cost adder.

Clearly, redundancy can enhance the reliability of a system, but with the addition of costs. As part of a DfR analysis, it would be necessary to calculate the life expectancy of the system structure to understand the impact of redundant systems.

3.7   Plating Materials: Tin Whiskers

Most of the electronics industry by now knows about tin whiskers, shown in Figure 3.18. They know whiskers are slim metallic filaments that emanate from the surface of tin platings. They know these filaments are conductive and can cause shorts across adjacent conductors. And they know that these shorts can cause some serious failures. But with all this knowledge, the industry is still struggling about how to predict

Figure 3.18
Tin whiskers.
Source:
Ansys-DfR
Solutions.

0.023″ (600 μm)

and prevent these "nefarious needles of pain" (Hillman et al. 2014). A good start to solving this problem is to admit that the formation of tin whiskers is not such a mystery. It does not arise due to bad karma, the black arts, or voodoo economics. Whiskering occurs because of the presence of a compressive stress (or, more accurately, a stress gradient). This compressive stress drives the preferential diffusion of tin atoms. A few more things then must occur for whiskers to form and grow, but in the absence of such stress, whiskering does not occur.

The issue to date has been the electronic industry's excessive focus on the indirect causes of whisker formation. The use environment (ambient temperature, thermal cycling, elevated humidity, and bending) simply initiates or accelerates the stresses present within the plating. It is the stress itself that should be modified, measured, and tracked over time to capture whisker behavior. The inability to capture and quantify the stresses present within the tin plating and to clearly delineate the sources of these stresses is stifling the eventual prevention (not just mitigation) of whisker-induced failures.

The stresses that drive whiskering derive from five sources:

1. Base metal (intermetallic formation)
2. Base metal (differences in coefficient of thermal expansion)
3. Bulk plating conditions
4. Oxidation/Corrosion
5. External pressure

Whiskering occurs when one or more of these sources induce stresses of enough magnitude. The magnitude of these stresses can be fixed at the time of production or can evolve over time. As a portion of the DfR assessment, it is often useful to do a tin whisker evaluation.

An analysis is conducted as follows. Perform an analysis of each of the methods for which compressive stress can be generated in tin plating, paying attention to the environment that the product will be subject to. Each component in the BOM is then evaluated for risk based on the method of stress generation. Follow-up work or information gathering may be required for those components deemed to be at-risk following the detailed analysis.

Because of significant differences in the diffusion rate of copper (Cu) through tin (Sn) grains and grain boundaries at room temperature, copper-tin intermetallic (Cu_6Sn_5) tends to grow preferentially into

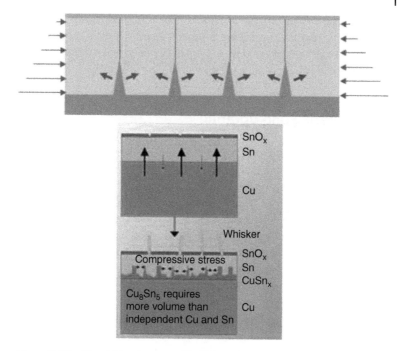

Figure 3.19 Tin whisker intermetallic formation.

grain boundaries. The volumetric expansion of Cu6Sn5 compared to Cu and Sn alone results in large compressive stresses within the plating (a diagram is shown in Figure 3.19). Annealing the tin coating immediately after plating at a temperature from 150–170 C is also commonly used to mitigate whisker growth. At temperatures over 60 C, the intermetallic that forms is Cu3Sn. In addition, at temperatures above 75 C, bulk and grain boundary diffusion rates in Sn become roughly equivalent. The resulting intermetallic morphology does not induce compressive stresses and provides a uniform layer that reduces the rate of diffusion and intermixing of Cu and Sn. It is essentially a poor man's Ni layer.

Based on the supplier part numbers provided in the BOM, datasheets are uploaded to get the spacing of the leads on the components, the base metal material, and the plating materials. The detailed BOM scrub data for unique parts that are used identifies those having package pin pitches less than 0.65 mm. Small passive components such as resistors and capacitors (0402 and smaller) can have termination caps that are spaced closer than 0.5 mm. Such components have been commonly tin plated for over a decade, and we are not aware of any reported cases of a tin whisker shorting out such a device.

3.8 Derating and Uprating

Component derating is the practice of reducing stresses on electronic components to improve product reliability. Several derating standards have been established by government and industry bodies. Examples include Air Force AFSC Pamphlet 800-27, MIL-HDBK-1547, NAVSEA SD-18, NASA Johnson Space Center SSP 30312, NASA Goddard MSFC-3012, NASA Jet Propulsion Lab JPL-D-8545, Boeing MF0004-400, and RIAC Quanterion. These derating approaches mostly improve product robustness when properly implemented (Kong 2013).

The starting point for component derating is often based on the manufacturer's datasheets or specifications. However, typical specifications provided by manufacturers are mostly functionality focused and often do not include critical reliability and durability parameters. Consequently, most derating parameters are limited to voltage, current, power, and temperature. In our view, stress evaluation in component derating can include additional considerations such as these:

- Types of stresses in addition to the electrical and operating temperatures, such as mechanical and chemical
- Attributes of stresses such as transient voltages and temperature
- Life-cycle consideration beyond operating stresses, such as manufacturing process effects
- Multi-stress effects (e.g. fretting corrosion, electrochemical migration)

The impact of derating should be evaluated scientifically when feasible, such as based on PoF knowledge. As an example, certain temperature-related derating rules implicitly assume that failure is always driven by elevated temperature and can be modeled by the Arrhenius equation. This is the case when derating the junction temperature for semiconductor devices. Reducing the junction temperature mostly helps device operational reliability, but there are questions worth asking when implementing the junction temperature derating rule, e.g.:

- What failure mechanisms are addressed by the junction temperature derating, and how is the activation energy obtained if the Arrhenius equation is used? For example, HCI in a CMOS structure is not accelerated by increasing temperatures.
- How is the derating helping with wearout life vs. overstress failure rates? These two attributes correspond to the different segments of a typical bathtub curve depicting reliability cycles, i.e. wearout region vs. intrinsic failure period. Reducing the semiconductor operating temperature typically both helps with high-temperature-related wearout performance (e.g. temperature-dependent dielectric breakdown and electromigration) and reduces thermal runoff possibilities, such as related to latch-up events.
- Junction temperature is a macro description, which can have a spatial and temporal distribution dependent on the device structure and operating parameters. Local hot spots can cause early failures, which is an increasing concern as devices scale down to ever smaller feature sizes.
- Is the generic derating rule applicable to the specific IC package and system it's embedded in? For example, system-in-package (SiP) could drive a different extent of derating to reach a specified performance than a QFN package style.
- Derating is not always "the more, the better," as different failure modes can emerge under different stress levels. HCI can become critical at reduced temperature levels, and a proper derating should be considered.
- Upon device scaling, leakage current also plays an important role. When leakage-related power consumption is no longer negligible, the junction temperature has an additional dependency on the leakage current, which is proportional to $e^{(-Ea/kT)}$, where Ea is the activation energy, k is the Boltzmann constant, and T is the temperature in Kelvin. In this case, the junction temperature can no longer be simplistically treated as only dependent on ambient temperature; thermal runoff can result from the temperature-activated leakage current increase.

Derating rules are not set in stone, and they should evolve with advances in component technologies in addition to any application-specific requirements and vendor capability assessments. As such, a derating assessment is a valuable element of the DfR process.

3.9   Reliability of New Packaging Technologies

The fabless semiconductor market is constantly searching for innovations to tackle advanced silicon technology. Large BGAs and QFNs are in development at various packaging companies. Substrates with complex copper traces and vias are constantly being pushed to tighter tolerances and smaller structures. The increased complexity is driving a decrease in reliability. There is a distinct advantage to creating more robust substrates, but it is quite difficult to figure out how to reduce problems that happen inside the package buildup without manufacturing and testing (Sharon et al. 2015).

Tools for quickly predicting whether a substrate design will meet reliability expectations are advancing as fast as the on-chip architecture. Finite element methods have emerged as the forefront of prediction and prevention of various failure modes and manufacturing challenges. In the last few years, the need for models with high-fidelity geometry has risen. Several schemes for refining meshes and smearing properties automatically are being used today. Creating high-fidelity models is still a labor-intensive endeavor for most finite-element modelers. Automatically generated, full three-dimensional models that include every geometrical feature are the next step. For this analysis, a 25 mm × 25 mm multi-chip module (MCM) was selected for assessment. A three-dimensional model of the substrate was created using the Sherlock tool. Every segment of geometry is modeled for every trace, and material properties are assigned automatically by the software. Figure 3.20 illustrates the detailed geometry modeled.

Package warpage is caused by copper imbalance between the two sides of the substrate stackup. Prediction of package warpage can be performed using effective properties and simpler models. The detailed model can provide information about localized effects of each via. Excessive warpage can hinder the creation of solder joints when a die is attached to the pads, and malformed solder joints can lead to a higher probability of cracks forming in the joints. Figure 3.21 shows how the analysis can identify specific issues down to the via level and point out how the stresses impact new package reliability.

This type of analysis can be performed on any new package development activity to ensure that the new package configuration will be capable of withstanding application stresses.

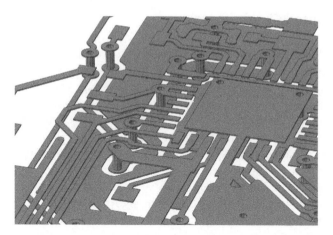

Figure 3.20 Detailed geometry and mesh of traces and vias.

Pad with low stress

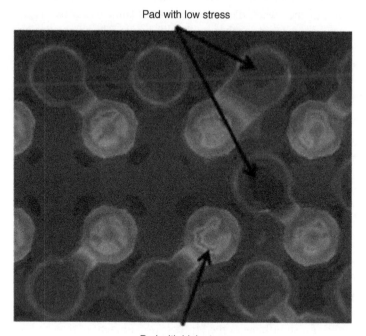

Pad with high stress

Figure 3.21 Detailed view of high- and low-stress pads. Source: Ansys-DfR Solutions.

3.10 Printed Circuit Boards

Designing PCBs today is more difficult than ever before due to significantly increased density, higher lead-free processing temperatures, and the associated changes required in manufacturing. The entire SUPPLY CHAIN has also been impacted by requirements regarding the use of hazardous materials and recycling. The RoHS and Registration, Evaluation, Authorisation, and Restriction of Chemicals (REACH) directives have caused suppliers throughout the industry to change both their materials and processes. So, everyone designing or producing electronics has been affected, even if they've been part of categories designated as directive-exempt. PCBs should always be considered a critical commodity. Without stringent controls in place for PCB supplier selection, qualification, and management, long-term product quality and reliability are neither achievable nor sustainable (Tulkoff et al. 2013).

A PCB is formed when dielectric material is laminated between layers of metal conductors or other conductive media. The conductive circuit is generally copper. Soldermask material insulates the outer layer conductive surface and prevents solder flow to areas that are masked or covered. Ascertaining the reliability of the PCB can be difficult, as it can be driven by:

- Size: Larger boards tend to experience higher temperatures.
- Thickness: Thicker boards experience more thermal stress.
- Material: Lower Tg tends to be more susceptible.
- Design: Higher density, higher aspect ratios.
- Number of reflow cycles.

For a proper DfR analyses, an understanding of the PCB processes and failure modes is beneficial.

3.10.1 Surface Finishes

The selection of the surface finish to be used on your PCBs could be the most important material decision made for the electronic assembly. The surface finish influences the process yield, amount of rework necessary, field failure rate, ability to test, scrap rate, and, of course, cost. One can be led astray by selecting the lowest-cost surface finish, only to find that the total cost is much higher. The selection of a should be done with a holistic approach that considers all important aspects of the assembly. Multiple surface finish options exist; each has its pros and cons that affect fabrication, solderability, testability, reliability, and shelf life. The five most utilized finishes are:

- Electroless nickel/immersion gold (ENIG) and ENEPIG (electroless Pd added)
- Immersion silver (ImAg)
- Immersion tin (ImSn)
- Organic solderability preservative (OSP)
- Lead-free hot air solder leveled (HASL)

Often, the approach for selecting the finish is an issue within companies. So, what is your selection approach?

- Component procurement: Select the cheapest one, and let the engineers figure out how to use it.
- PCB engineer: Select the finish that is easiest for the suppliers to provide (their sweet spot); let the assembler figure out how to use it.
- Assembly engineer: Select the finish that provides the largest process window for assembly and test.
- Sustaining engineer: Select the finish that minimizes field failures.
- CEO: Select the finish that minimizes the overall cost (including reliability risk).

There are several things to consider when dealing with surface-finish selection:

- Cost sensitivity
- Volume of product (finish availability)
- SnPb or LF process
- Shock/Drop a concern?
- Cosmetics a concern?
- User environment (corrosion a concern)?
- Fine pitch assembly (less than 0.5 mm)?
- Wave solder required (PCB thickness greater than 0.062 in.?
- High-yield ICT important?

Next, the pros and cons of each approach are discussed so you can perform your DfR analysis properly.

3.10.1.1   Organic Solderability Preservative (OSP)

OSP is an organic compound that bonds to the copper on the surface of the PCB and forms a layer that protects the copper before being subjected to a soldering operation. It provides a flat surface, is a no-lead process, and is relatively inexpensive. Some issues with OSP are that it has a lower wetting force, which can lead to insufficient hole fill with solder. This can have a significant impact on through-hole parts being properly soldered. OSP also has a short shelf life, which can impact storage of PCBs and how quickly they must be processed once introduced to the manufacturing line. Of importance is that OSP, because it is a barrier, can be an issue when in-circuit-testing is performed. The probes do not penetrate the coating well, leading to incorrect test results.

3.10.1.2   Immersion Silver (ImAg)

ImAg is a single-material system specified by IPC-4553. It has good flatness and co-planarity, has a longer shelf life when stored properly, has good wettability and good testability, and is low cost. However, there are several issues with this finish. Galvanic etching can occur at the solder-mask edge if the PCB rinsing and drying operations are not properly performed during fabrication. This etching can cause total failure electronically. Figure 3.22 illustrates this failure mechanism.

ImAg is also susceptible to surface tarnish creating a solderability issue, as there is significant degradation in the solderability. For example, this can occur in less than one day in a mixed-flowing gas test that is equivalent to six months in the field in a light industrial environment.

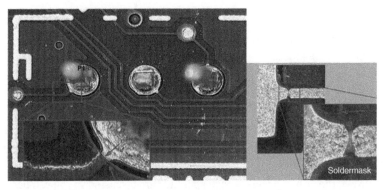

Figure 3.22   Immersion silver galvanic etching. Source: Ansys-DfR Solutions.

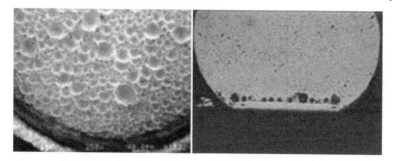

Figure 3.23 Champagne voiding. Source: Ansys-DfR Solutions.

Another issue involves champagne voiding, where many small voids occur at the solder/plating interface. The root cause of this issue is still not well understood. Figure 3.23 illustrates this failure mechanism.

However, the most problematic issue with ImAg is its susceptibility to corrosion. The first is creeping corrosion, where the resistance to dendritic growth reduces as humidity increases. Circuit traces susceptible to leakage current trigger the failure. Figure 3.24 illustrates this failure mechanism. In addition, ImAg is impacted by sulfur in the air. Sulfur interacts with the silver and results in dendrites growing rapidly on the surface of the PCB.

These corrosion failures are driven by the environment where your product might be in operation. These environments must be considered as part of your DfR analysis:

- Paper mills
- Rubber manufacturing (tires, for example)

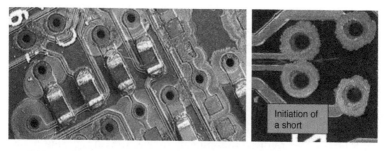

Figure 3.24 Creep corrosion. Source: Ansys-DfR Solutions.

- Fertilizer
- Wastewater treatment
- Mining/smelting
- Petrochemical
- Clay modeling studios
- Etc.: Includes companies nearby such industries

3.10.1.3 Immersion Tin (ImSn)

ImSn is specified by IPC-4554. The standard thickness is one micron (40 microinches), with some companies specifying up to 1.5 microns. This finish also has excellent flatness and is low cost. It is not as popular due to the environmental and health concerns regarding thiourea, a known carcinogen. There is also a concern with tin whiskering. There are, however, several issues with this surface finish. If the finish is too thin, then there can be a solderability problem after storage or after a second reflow cycle due to an intermetallic (IMC) growth through the thickness. Solderability problems can also occur due to an oxide buildup on the surface that is greater than 5 nm. Exposure to humidity (greater than 75%) greatly accelerates the oxide growth through the creation of tin hydroxides. The use of sealed moisture and airtight wrapping for shipment helps to mitigate this issue. Finally, cleanliness of the PCB is also a possible issue, as contamination breaks down the self-limiting nature of tin oxides.

3.10.1.4 Electroless Nickel Immersion Gold (ENIG)

The ENIG finish is a two-material system specified by IPC-4552. The electroless nickel thickness should be 3–6 microns, with a thin gold top coat 0.08–0.23 microns thick for optimum reliability. Like many of the other finishes, ENIG has excellent flatness and long-term shelf life. It also has excellent oxidation resistance and wettability and is robust for multiple reflow soldering cycles. The gold finish readily dissolves into solder and does not tarnish or oxidize. Like the others, ENIG also has its issues. In the past, a failure mode called black pad was a major issue with this surface finish. It is associated with the phosphorous content: if the content was too high, there were weak phosphorous-rich areas after soldering; and if the level was low, the insufficient phosphorous did not prevent corrosion during the highly acidic immersion gold process, causing black pad. US suppliers resolved this issue more than 10 years ago, but it has re-emerged with newer Chinese suppliers. As such, if you procure your circuit boards domestically, you should not have a problem; but if you buy from sources in China, you must be aware of the potential problem. It is recommended that you perform solderability testing on samples from each lot of boards to assure that this problem does not occur. ENIG is also not as strong as other finishes with respect to mechanical shock and vibration. ENEPIG, which introduces a layer of palladium into the stackup, has no issues with black pad due to the palladium barrier. However, it is a much more costly surface finish.

3.10.1.5 Lead-Free Hot Air Solder Leveled (HASL)
The HASL finish has excellent solderability and wettability. It is also excellent with respect to shelf life as well as being robust in multiple reflow cycles. As a result, joints are solder to solder, which produces strong solder joints. Its drawbacks are that it is difficult to maintain co-planarity, and there is a risk of a fabrication issue with the circuit board from the process of dipping the board into the hot solder liquid.

Table 3.2 summarizes some of the challenges of using the LF HASL finish.

To accommodate the co-planarity issue, the recommended thickness of the HASL finish should be 4 microns (100 microinches). Lower minimums can result in exposed intermetallic. One manufacturability issue is thickness variability among the various pad sizes (footprints)

Table 3.2 Lead-free HASL challenges.

Challenge	Mitigating action
Heat damage to board: LF HASL is run at 265 C vs. 255 C for SnPb HASL	Change laminate from dicy to phenolic (lower z-axis expansion)
	Increase copper thickness in vias
Limited availability	Work closely with PCB suppliers to add necessary capacity
Planarity of finish	Select the optimum alloy (high fluidity)
	Optimize process parameters to control planarity (bath temp, air knives, etc.
Copper dissolution	Select the optimum alloy
	Preheat assemblies to rework

used in the surface-mount technology (SMT) process. You might see 100 microinches over a set of small BGA pads, but over 1,000 microinches on large pads. This is typically controlled through a hot air knife process during fabrication, where the air knife pressure, solder pot temperature, and nickel content are critical.

Different finishes tend to be the best fit in certain situations:

OSP (must address the ICT issue) is good for:
- Hand-held electronics
- Notebook computers
- Basic desktop computers
- Basic consumer electronics and power supplies
- LF medical or aerospace (thin PCBs)

ENIG and ENEPIG are good for medical and aerospace applications and small specialty electronics where drop shock is not a major environment. ImAg is good in fully enclosed electronics, RF applications, and some basic (but sealed) consumer electronics. ImSn is suitable for simple consumer electronics (not fully enclosed), simple medical or aerospace applications (single-sided), and low- to moderate-volume peripheral components. LF HASL is compatible with thick LF PCBs in office/telecom environments and complex medical and aerospace applications.

What happens if there is no good fit for a surface finish? For example, suppose you need low cost, high volume, and corrosion resistance, with a good ICT capability circuit board. One solution might be to use ImAg but plug vias with soldermask to protect from corrosion (but cost is sacrificed). Another is to use OSP but implement cleaning to remove flux residue for probing (cost is sacrificed). Although no perfect solution exists, the assemblies can be produced, albeit at a cost impact. The surface finish you select will have a large influence on quality, reliability, and cost. It is a complex decision that impacts many areas of the business. Select a finish that is optimal for the business (and not just one function). Know that there are engineering tricks to improve on weak areas of each finish. Stay current in this field, because new developments continue to be made. And be aware that the DfR analysis can help to sort out this type of situation.

3.10.2 Laminate Selection

Table 3.3 provides insight into a methodology for selecting the right laminate for your application. In addition to the table, note:

1. If copper thickness = 2 oz, use the material listed in column 260 C.
2. If copper thickness is greater than or equal to 3 oz, use phenolic base material or high-Tg halogen-free materials only.
3. Twice-laminated product: Use phenolic material or high-Tg halogen free materials only (includes HDI).
4. Follow customer requirements if the customer specifies materials.
5. Confirm the IR reflow temperature profile.

The thermal properties of laminate material are primarily defined by four parameters, with each capturing a different material behavior.

- Out-of-plane coefficient of thermal expansion (Z-CTE)
- Glass transition temperature (Tg)
- Time to delamination (T260, T280, T288)
- Temperature of decomposition (Td)

Let's look at those parameters in a little more depth

- Out-of-plane CTE (below Tg or Z-axis: 50 C to 260 C): CTE for SnPb is 50–90 ppm (50 C to 260 C rarely considered). LF: 30–65 ppm
- Glass transition temperature (IPC-TM-650): Characterizes complex material transformation (increase in CTE, decrease in modulus). Tg of 110 C to 170 C for SnPb. LF: 150 C to 190 C.
- Time to delamination (IPC-TM-650, 2.4.24.1): Characterizes interfacial adhesion. T-260 for SnPb is 5–10 minutes. LF: T-280 of 5–10 minutes or T-288 of 3–6 minutes.
- Temperature of decomposition (IPC-TM-650, 2.3.40): Characterizes breakdown of epoxy material. Td of 300 C for SnPb. LF: Td of 320 C.

The appropriate material selection is driven by the failure mechanism you are trying to prevent (e.g. cracking and delamination, PTH fatigue, conductive anodic filament formation).

3.10.3 Cracking and Delamination

Fiber/resin interface delamination occurs as a result of stresses generated under thermal cycling due to a large CTE mismatch between the

Table 3.3 Laminate material selection.

Board thickness	IR-240~250 C	Board thickness	IR~260 C
<60 mils	Tg 140 dicy All HF materials OK	<60 mils	Tg 150 dicy HF- middle and high Tg materials OK
60~73 mil	Tg 150 dicy NP150, TU622-5 All HF materials OK	60~73 mil	Tg 170 dicy HF- middle and high Tg materials OK
73~93 mil	Tg 170 dicy, NP150G-HF HF- middle and high Tg materials OK	73~93 mil	Tg 150 phenolic + filler IS400, IT150M, TU722-5, GA150 HF- middle and high Tg materials OK
93~120 mil	Tg 150 phenolic + filler IS400, IT150M, TU722-5 Tg150 HF- middle and high Tg materials OK	93~120 mil	Tg 170 – phenolic IS410, IT180, PLC-FR370 Turbo, TU722-7 HF- middle and high Tg materials OK
121~160 mil	Tg 170 – phenolic IS410, IT180, PLC-FR370 Turbo, TU722-7 HF – high-Tg materials OK	121~160 mil	Tg 170, phenolic + filler IS415, 370HR, 370 MOD, N4000-11 HF – High-Tg materials OK
>161 mil	Tg 170, phenolic + filler IS415, 370HR, 370 MOD, N4000-11 HF material TBD	>161 mil	TBD – consult engineering for specific design review

glass fiber and the epoxy resin (1 vs. 12 ppm/C). Delamination can be prevented/resisted by selecting resin with lower CTEs and optimizing the glass surface finish. Studies have shown that the bond between fiber and resin is strongly dependent on the fiber finish. Morphology and location of the cracking and delamination can vary, even within the same board. Failure morphology and locations occur within the middle and edge of the PCB; within prepregs and/or laminate; and within the weave, along the weave, or at the copper/epoxy interface (adhesive and cohesive).

How do you home in on the cracking or delamination issue? Some PCB suppliers with a non-optimized process have demonstrated improvement through modifications to the lamination process or oxide chemistry. They also have lot-to-lot variability. Others encounter limits to their fabrication capabilities, as it is difficult to overcome the adhesion vs. thermal performance tradeoff (dicy vs. phenolic), or the high stresses developed during LF fabrication exceed the material strength of standard board material. Moisture absorption can also be a contributor.

As part of a DfR analysis, it may be necessary to audit the fabrication facility to ascertain how certain operations can translate into moisture absorption issues. These may be:

- Storage of prepregs and laminates.
- Drilling process: Moisture is absorbed by the side walls (microcracks?) and trapped after plating.
- Storage of PCBs at the PCB manufacturer.
- Storage of PCBs at the CCA manufacturer.

Delamination of the traces from the surface of the circuit board can also occur. Excessive temperatures during high-temperature processes, insufficient curing of resin, and insufficient curing of soldermask are sources of increased stress. Conversely, sources of decreased strength include improper preparation of copper foil and excessive undercut in fabrication.

3.10.4   Plated Through-Holes and Vias

Several issues can affect PTH reliability. Voids can cause large stress concentrations, resulting in crack initiation. The location of the voids can provide crucial information in identifying the defective process: around the glass bundles, in the area of the resin, at the inner layer interconnects (aka wedge voids), and at the center or edges of the PTH. Etch pits are also an issue. They are due to either insufficient tin resist deposition or improper outer-layer etching and rework. These etch pits can cause large stress concentrations locally, increasing the likelihood of crack initiation, which can result in an electrical open.

Overstress cracking of a PTH is typically due to a CTE mismatch that places the PTH in compression. Because ICT is rarely performed at operational temperatures, it can drive a localized deflection of the PCB, resulting in solder joint cracks and barrel cracks in the circuit board. This is illustrated in Figure 3.25.

The most common issue with PTHs is fatigue of the barrel, causing a crack to initiate and ultimately causing the PTH to fail. This is caused by circumferential cracking of the copper plating that forms the PTH wall. It is driven by differential expansion between the copper plating (≈ 17 ppm) and the out-of-plane CTE of the printed board (≈ 70 ppm). When a PCB experiences thermal cycling, the expansion/contraction in the z-direction is much greater than that in the x-y plane. The glass fibers constrain the board in the x-y plane but not through the thickness. As a result, a great deal of stress can be built up in the copper via barrels, resulting in

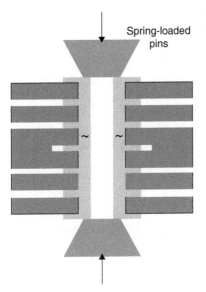

Spring-loaded pins

Figure 3.25 Compression on a PTH from ICT.

Figure 3.26 PTH barrel crack. Source: Ansys-DfR Solutions.

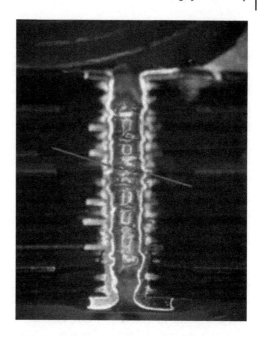

eventual cracking near the center of the barrel. The industry-accepted failure model is IPC-TR-579. Figure 3.26 shows a barrel crack.

So, from a DfR analysis perspective, several things can be accomplished to help ensure a robust PCB. These include:

- Reduce laminate/prepreg thickness (high-technology boards are routinely fabricated with laminate/prepreg thicknesses of less than 100 microns). This redesign tends to result in minimal cost changes and have a minimal effect on other aspects of the design but has the least effect on PTH reliability.
- Specify thicker plating on the PTH wall. The IPC minimum is 20-25 microns.
- Specify up to 40 micron thickness (a larger value results in knee cracks). This redesign will also have a minimal cost but strongly depends on the quality of the board fabricator (adjusting the process may result in the introduction of other types of defects).
- Another option for improved reliability is to change the board material to a laminate that has a higher Tg (greater than 170 C) and Td (greater than 340 C) as well as an out-of-plane CTE of less than 70ppm.

From a reliability standpoint, you can also work to maintain a low aspect ratio of the PCB thickness to the drill hole diameter. It is recommended to keep the ratio below 10. For example, on a 93-mil-thick board, the minimum hole size should be nine mils. Also, as most field failures have been observed to be due to drilling, smearing, and plating defects, it is recommended to perform strong statistical process control (SPC) of these processes. Finally, using the appropriate coupons for fabrication provides tools for assessing the PTHs, and the coupons can be monitored both at the fabricator as well as at your facility.

3.10.5 Conductive Anodic Filament

Conductive anodic filament (CAF), also referred to as metallic electromigration, is an electrochemical process that involves the transport (usually ionic) of a metal across a nonmetallic medium under the influence of an applied electric field. CAF can cause current leakage, intermittent electrical shorts, and dielectric breakdown between conductors in printed wiring boards. Figure 3.27 illustrates an example where the four elements that can result in CAF came together: an internal fault in the circuit board, an electrolyte, bias, and moisture. When those four factors occur,

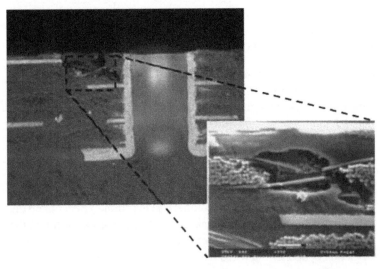

Figure 3.27 Conductive anodic filament formation. Source: Ansys-DfR Solutions.

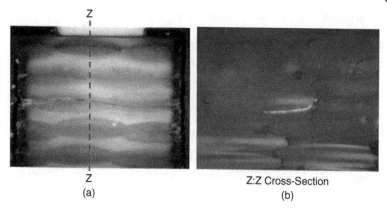

Figure 3.28 Conductive anodic filament example. Source: Ansys-DfR Solutions.

a copper growth can form, migrate from one voltage extreme to another (or ground), and result in a short. Figure 3.28 shows another example.

The photo in Figure 3.28(a) shows a cross-sectional view of conductive anodic filaments between two PTHs. An oblique slice through the copper filament is shown in Figure 3.28(b).

Classic CAF occurs along the fiber/epoxy interface, while exposure to elevated temperature-humidity conditions weakens glass/polymer bonds based on silanes.

There is some debate about critical paths. PTH-to-PTH has the potential to create the greatest internal damage due to the larger exposed surface area. If the spacing between the PTHs is greater than 20 mils, there should not be a problem. The risk increases as the spacing decreases. PTH to plane has less potential for damage and requires a 7–10 mil clearance to ensure that an issue does not occur.

There are several failure accelerators of CAF. Those that cause a decrease in strength are listed here:

- Fiber/resin separation
- Copper/resin separation
- Hollow fibers
- Poor wet out
- Misregistration of PTHs/vias
- Copper wicking
- Drilling damage

- Separation of PTH wall from epoxy resin

The following issues increase stress:

- Flux chemistry
- Higher voltage
- Higher moisture concentration
- Tighter conductor spacing
- Higher ion concentration in epoxy resin

Fiber/resin interface delamination occurs as a result of stresses generated under thermal cycling due to a large CTE mismatch between the glass fiber and the epoxy resin (ratio of 1 to 12). Delamination can be prevented/resisted by selecting resin with lower CTEs and optimizing the glass surface finish. Studies have shown that the bond between fiber and resin is strongly dependent on the fiber finish.

Hollow fibers, which form from decomposed impurities in the glass melt, are an alternate path for CAF. Figure 3.29 shows an example of hollow fibers.

Generally, CAF is a two-step process that is dependent on debonding between the glass fibers and epoxy resin matrix to provide a path for copper migration. With the appearance of hollow fibers inside the laminates, CAF can happen as a one-step process. As such, any concentration of hollow fibers in laminate becomes critical to reliability.

The PCB industry has responded to these issues by creating CAF-resistant laminates that use different binders in epoxy formulation.

Figure 3.29
Hollow fiber
example.
Source:
Ansys-DfR
Solutions.

This approach is still expensive. An alternate, though not proven, is the use of phenolic cured epoxy. Similarly, there have also been changes to the weave of glass cloth and improvements in silane application.

High-layer-count PCBs combined with thickness restrictions result in single-ply core/laminates. Single-ply laminates are sensitive to conductive anodic filaments at voltages below 5 volts. Therefore, some companies specify dual-ply or a minimum thickness of 3 mil.

Another form of CAF is vertical filament formation (VFF). VFF is a CAF-type mechanism that results in vertical migration of conductive filaments between power and ground. This VFF is especially insidious because it can occur in low-voltage (3.3 V) products operating in benign environments (such as offices). Spacing between power and ground planes should be less than the mesh used to sieve the liquid epoxy and should not cause nesting.

As part of a DfR design review:

- Ensure susceptibility first (may require testing).
- Use CAF-resistant grade laminate (available from Isola, Polyclad, Matsushita, etc.) Can be dependent on fiber weave, fiber coating, or epoxy material.
- Increase the Td and/or T288 of laminate material.
- Move to phenolic-cured resin.
- Reduce the usage of non-plated through-holes.
- Increase the distance between adjacent PTHs.
- Increase the spacing between PTH and power and ground planes.

3.10.6   Strain and Flexure Issues

Excessive flexure of the circuit board can occur during several post-assembly process operations. For example, flexure can happen during the depanelization operation, in-circuit testing, installation into a housing or fixture, connector installation (press fit), heatsink installation, etc. Board flexure can also be impacted by mechanical shock and vibration stresses.

The strain level noted during any of these operations should not exceed 500–750 microstrain. At this point, the potential for cracking of the solder joints or components themselves goes up dramatically.

Calculating the PCB-level strain is more complex. Except for simple structures, you need finite element analysis (FEA). There are techniques that use simple spring mass approximation to predict the board deflection during a shock event. Spring/mass models assume masses connected by ideal weightless springs. However, FEA simulations are usually transient dynamic in nature. As such, ANSYS's Sherlock tool uses an implicit transient dynamic simulation (useful when solving linear/elastic) issues. Shock pulses are transmitted through the mounting points into the circuit board. The resulting strains are extracted from the FEA results and used to predict robustness under shock conditions. Figure 3.30 illustrates the strain level on a PCB when subjected to a 50 G shock pulse causing a 12 mm deflection, which is severe.

Performing these analyses as part of the DfR assessment is extremely beneficial to the design process. Figure 3.31 shows another example of the strain levels on a PCB causing excessive strain and deflection.

During the analysis, two additional mounting points were added, midspan, to the circuit board. This resulted in the deflection dropping from 12 mm to 1.65 mm, as shown in Figure 3.32. There are still failures, so additional support is required. The key is that these analyses can be performed in minutes so a viable solution can be ascertained.

Besides modeling, what other actions can you, as the reliability engineer, take to reduce strain levels associated with mechanical shock stresses? As shown, the addition of mount points provides support.

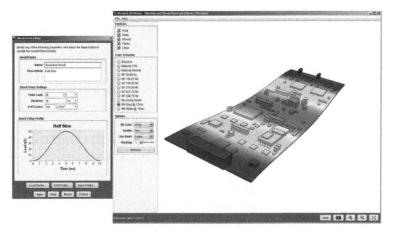

Figure 3.30 Strain level from a 50 G mechanical shock. Source: Ansys-DfR Solutions.

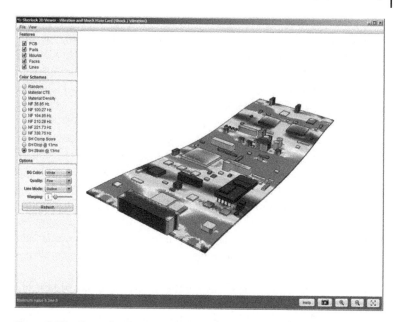

Figure 3.31 Example of excessive strain. Source: Ansys-DfR Solutions.

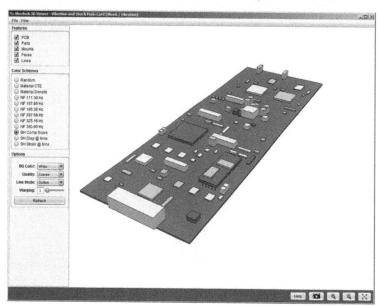

Figure 3.32 Reduced strain after mount points are added. Source: Ansys-DfR Solutions.

Standoffs, bonding the components to the PCB, and using a thicker circuit board are also ways to enhance mechanical strength. Flexible termination capacitors are another option for reducing the effect of flexure. Similarly, you can use corner staking, edge bonding, or underfill.

ANSYS-DfR performed a study that addressed the issue of which technique offered the best mitigation. This study focused on 208 I/O BGAs on a four-layer FR4 board. Three solders were tested: two LF (SAC305 and SN100C) and one leaded (SnPb). Three mitigation techniques were tested: corner staking, edge bonding, and underfilling. For each of these techniques, two mitigation materials were tested. One material was reworkable and the other was not. The boards were subjected to mechanical shock testing and sinusoidal vibration testing until failure. The results of the testing show that no one mitigation technique is best for all the conditions tested. The same is true for the mitigation material. The best choices of mitigation technique and material are application-dependent.

With the enforcement of RoHS requirements, there has been some concern about the reliability of LF solders. The reliability of SnPb solder as a robust solder is well known, and the failure rate is well characterized, whereas tin-silver-copper (SAC) solders are known to be less robust, which is a problem for some high-reliability applications. One solution is to mitigate the solder joints with a staking material to strengthen them, but the problem is choosing the correct material for the application. Several material properties, such as Tg and CTE, need to be properly chosen. Improperly chosen mitigation materials will cause premature failures instead of extending the time to failure. If the expected operating temperature of the device is too close to the Tg, the CTE will drastically change, which can cause solder joint failures. Figure 3.33 shows the difference in application for corner staking, edge bonding, and underfill used in the experiments.

Figure 3.33 Corner staking, edge bonding, and underfill. Source: Ansys-DfR Solutions.

Testing included drop-shock testing at 750 G (25 C), 1,000 G (25 C), and 1,500 G (-55 C, 25 C, and 125 C). The sinusoidal vibration was performed at 90 mils displacement and the same three temperature levels. The random vibration was performed using PSD levels of 0.1 G2/Hz at 25 C. The results again indicated that the different techniques for mitigation depended on the specific application (Caswell and Keener 2016).

The point of this discussion is that testing is often needed to confirm the best approach for either a test plan or a mitigation technique. It needs to be part of the arrows in the quiver for reliability analysis.

3.10.7   Pad Cratering

Pad cratering is a crack initiating in the laminate during a dynamic mechanical event. It could be from ICT testing, depanelization, connector insertion, mechanical shock (being dropped), or vibration. The drivers for this failure mode are finer-pitch components and more brittle laminates for LF assembly. It is difficult to detect using standard procedures (e.g. X-ray, dye and pry, ball shear, and ball pull). Figure 3.34 illustrates the ways pad cratering can manifest, either as a no-failure or open circuit.

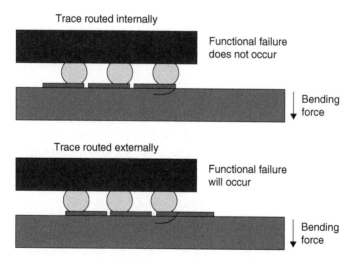

Figure 3.34   Pad cratering cross section.

There are some things you can do as an element of the DfR assessment to mitigate the potential for pad cratering:

- May require additional characterization (e.g. high-speed pull test).
- May require redesign (solder-mask-defined vs. non-solder-mask-defined pads).
- May require limitations on board flexure (750 to 500 microstrain).
- Component-dependent (may require new LF solder). SAC305 is relatively rigid; SAC105 and SnNiCu are possible alternatives.

3.10.8   PCB Buckling

PCB buckling occurs due to excessive warpage of the PCB. The sources of increased stress are:

- Non-parallel lamination fixturing/press platens
- Absorption of moisture before assembly
- Excessive temperatures during assembly
- Rapid heating or rapid cooling during assembly
- Bad fixturing during wave soldering

There are also causes for decreased strength in the laminate.

- Incomplete cure of epoxy resin
- Asymmetrical stackup of board layers
- Unbalanced circuitry layers
- Inadequate post-lamination stress relief

3.10.9   Electrochemical Migration

Electrochemical migration (ECM) is defined, per IPC-TR-476A, "Electrochemical Migration: Electrically Induced Failures in Printed Wiring Assemblies," as "The growth of conductive metal filaments or dendrites on or through a printed board under the influence of a DC voltage bias." This definition is a little narrow. We think this is more appropriate: "Electrochemical migration is the movement of metal through an electrolytic solution under an applied electric field between insulated conductors." ECM can occur on or in almost all electronic packaging, at the die surface, in the epoxy encapsulant, on the PCB, on passive components, etc.

Field returns are often identified as ECM failures, as are many identified as "no trouble found." The self-healing mechanism makes ECM difficult to identify. Constant reductions in pitch between conductors makes current designs more susceptible to ECM. The increased use of leadless packages (BGAs, flip-chip devices) results in a reduction in standoff, reduces the efficiency of cleaning and may lead to increased concentration of contaminants. The frequency of ECM is expected to increase as electronic equipment makes inroads into tropical environments (Southeast Asia, Central America, etc.) because ECM occurrence is very sensitive to ambient humidity conditions. Finally, LF applications may require more aggressive flux formulations.

ECM is primarily due to the presence of contaminants on the surface of the board, with the strongest drivers being halides (chlorides and bromides). Weak organic acids (WOAs) and polyglycols can also lead to drops in the surface insulation resistance. As such, ECM is primarily mitigated through controls on cleanliness.

The following major factors affect electrochemical migration.

3.10.9.1 Temperature

The temperature range for ECM to initiate is between $0\,C$ and $100\,C$, assuming the electrolytic solution is aqueous. Outside this temperature range, other types of migration behavior that require no electrolytic solution may occur, such as the migration of silver at $150\,C$ (Benson et al. 1993). Studies have demonstrated that the propagation rate of dendrites tends to increase as a function of temperature (Lall et al. 1997; DiGiacomo 1982). Temperature has been associated with ECM through the chemical reaction rate defined by the Arrhenius equation. Because the initiation of ECM is a chemical oxidation-reduction reaction, it follows that the rate of initiation may have an Arrhenius dependence. While this dependence has been established in other corrosion-based failure mechanisms that afflict electronics, such as aluminum corrosion on a silicon die, it has not been demonstrated conclusively for ECM on PCB assemblies (PCBAs).

3.10.9.2 Relative Humidity

Relative humidity (RH) is the ratio of the mole fraction of water vapor in the air at a specific temperature to the maximum amount the air could hold at that temperature, expressed as a percentage. Water vapor can attach to the PCB by hydrogen bonding, providing the electrolytic solution that is needed for ECM to occur (Hallberg and Peck 1991). Once the RH is high enough for water to adsorb to the PCB, a critical moisture level is required on the surface in order for ECM to occur. This water level is expressed as number of monolayers of water. This critical layer of water is reported to be approximately 20 monolayers thick (Zamanzadeh et al. 1990). The relationship between the critical number of monolayers necessary for migration to occur and the ambient relative humidity is dependent on the moisture adsorption characteristics of the PCB surface. This adsorption behavior is influenced by the choice of soldermask and the quantity and type of contamination present on the surface. Usually, the higher the relative humidity, the faster the onset of initiation of ECM (Krumbein 1988). The dew point temperature is the temperature where water vapor may condense on the surface is a film, providing bulk liquid formation and almost instantaneous path formation. If the temperature drops below the dew point, the time for path formation will be much shorter than the time for the other stages of ECM, and the time to failure might be greatly reduced. For example, water drop tests can demonstrate dendritic growth in seconds; non-condensing ECM tests require hours before failure occurs.

3.10.9.3 Voltage Bias

Voltage bias must be present for ECM to occur. *Voltage bias* is the difference in voltage between the anode and cathode. This electric field drives ionic migration by causing the positive ions to travel along the field lines from the anode to the cathode through an ion-transport path provided by the aqueous medium (Rudra and Jennings 1995). A higher bias increases the rate of propagation and may increase the chance of catastrophic failure (i.e. fire, total loss of assembly).

3.10.9.4 Conductor Spacing

Conductor spacing refers to the distance between two oppositely charged conductors. As this distance decreases, time to failure will decrease if the rate of dendrite propagation stays constant. Other factors can also increase the frequency of failure with finer-pitch spacings. Smaller spacings tend to be more difficult to clean, potentially resulting in a buildup of contaminants. Smaller spacings at the same voltage also result in a higher electric field.

Some ECM mechanisms have more definitive descriptions: for example, dendritic growth is defined as a tree-like or feather-like growth on the surface of a PCB, as shown in Figure 3.35 (top). Conductive anodic filament (discussed elsewhere in this chapter) is described as a migration within the PCB. An example of CAF is also shown in Figure 3.35 (bottom).

Traditional ECM migration involves four steps:

1. Path formation
2. Electrodissolution
3. Ionic migration
4. Electrodeposition

The first step, path formation (Caswell and Binfield 2017), is the creation of an electrolytic solution that is sufficiently conductive. This means enough water exists on the surface and that the water is

Figure 3.35
Examples: dendrite (top) and CAF (bottom). Source: Ansys-DfR Solutions.

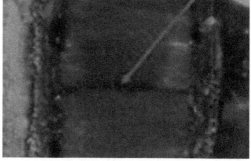

sufficiently conductive. The requirement for path formation is why electrochemical migration tends to be a "it happens or it doesn't" type of failure mechanism. If there is not enough water (relative humidity is too low) or the water is not sufficiently conductive (the PCB is clean), then ECM cannot occur. Both cations and anions can play a role in the formation of conductive electrolytic solutions. It is believed to be the rate-limiting step. Dendritic growth involves the creation of an electrolytic solution that is sufficiently conductive, often driven by relative humidity, contaminants, and delamination; and CAF involves a degradation of the epoxy/glass interface.

The second step, electrodissolution, is where the roles of anions and cations begin to diverge. *Electrodissolution* is the process of dissolving metals through electrolysis. Specifically, copper at the anode is reduced to Cu^{2+} and e^-. This dissolution step provides the copper ions that then migrate to the cathode, where the copper ions plate out as copper dendrites (the copper plates preferentially along closed packed crystallographic planes, which creates the dendritic shape). Dissolution requires a polar liquid (usually water) and an electric field with a voltage higher than the anodic electromagnetic force (EMF). In the case of copper, this is 0.34 V. Thus, copper can dissolve into water with no ions (deionized [DI] water) if the applied voltage is sufficiently high. The electrochemical reaction usually occurs adjacent to the anode, where an increased concentration of hydrogen ions reduces the pH below 7.0. The dissolution of the copper anode material then occurs. Halides are not required to induce dissolution but can increase susceptibility through changes in ion formation. The reduction in pH increases the solubility of metal ions (i.e. copper). Figure 3.36 illustrates this concept.

Ionic migration is also known as *electrophoresis* which is the migration of charged particles through a solution under the influence of an electric field. Positive ions travel along the field lines from anode to cathode, while electrons travel in the reverse direction.

Electrodeposition is a reaction that occurs once ions reach the cathodes, as shown in the flow equations.

Electrodeposition Flow

$$O_2 + 2H_2O + 4e^- \rightarrow 4OH^- \tag{3.1}$$

$$2H_2O + 2e^- \rightarrow 2OH^- + H_2 \tag{3.2}$$

$$Cu^+ + e^- \rightarrow Cu \text{ or } Cu^{2+} + 2e^- \rightarrow Cu \tag{3.3}$$

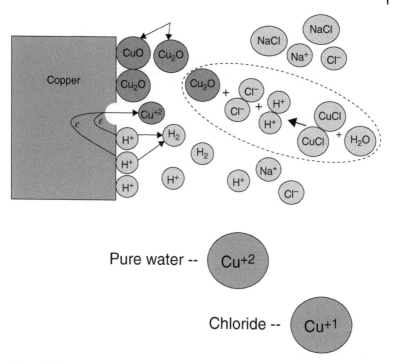

Figure 3.36 Electrodissolution.

Because cathodic electrode position is largely diffusion-controlled, the rate depends on the metal ion concentration in the aqueous medium. Production of hydroxyl ions (OH-) at the cathode can reduce the rate of electrode position by combining metal ions with hydroxyl ions to form hydroxides. Hydroxides remove metal ions from the aqueous medium through partial precipitation, and the ECM rate is correlated with the solubility product of the metal ion hydroxide. Ionic contaminants, such as Cl -, can further change the ECM rate by forming alternate reaction paths and additional ionic species: for example, CuCl in the case of copper and Cl-. The following are several drivers of ECM:

- Electrolytic solution
- Moisture: Temperature, humidity, cracks, contamination
- Ions: Contamination, cleanliness
- Electric field: Voltage, spacing

- Insulation: Soldermask, encapsulant, ceramic
- Conductors: Metal alloy

Several things must be considered as part of a DfR review/analysis regarding ECM. To understand an electrolytic solution, you must ask the following questions:

- Where does the water come from? Ambient moisture in the air, or evaporation of absorbed moisture (surface ECM)?
- What measurement (techniques) are available? Adsorption (quartz crystal, ellipsometry)? Absorption (weight gain)?
- What are the measurement units?
  - Adsorption: Monolayers of moisture or areal mass density (ng/cm^2). One monolayer = 31 ng/cm^2.
  - Absorption: Percent change in weight
- How much water? This is very dependent on relative humidity, yet relatively insensitive to temperature, even for moisture absorption (numerous internal interfaces, faster diffusion). Monolayers over polymer are difficult to measure because of the dueling processes of adsorption and absorption, which are influenced by surface properties (hydrophilic, hydrophobic), bulk properties, and porosity – parameters rarely measured by the materials supplier. The actual number of monolayers over a polymeric surface, as a function of relative humidity, is debatable, but is estimated to be a few to several hundred.

Some general guidelines for ECM risk are shown in Table 3.4.

Deliquescence is the dissolution of ionizable residues. When this occurs, resistance change can be several orders of magnitude (der Marderosian 1977). Each organic compound has a different critical RH. This is shown in Table 3.5.

Table 3.4 ECM risk guidelines.

Relative humidity	ECM likelihood
<60%	Rare
60 to 80%	Unlikely
>80%	Possible
Condensation	Likely

Table 3.5 Deliquescence characteristics.

Compound	Temperature(0 C)	%RH
LiCl.H20	20	15.0
KF	100	22.9
NaBr	100	22.9
CaCl2.6H20	24.5	31.0
CaCl2.6H20	5	39.8
KBr	100	69.2
NaCl	20	75.0
KCl	80	78.9
KBr	20	84.0
KCl	0	88.6
NaF	100	96.6

3.10.9.5 Condensation

What is condensation?

- When surface moisture becomes visible?
- The amount of adsorbed moisture at 100% RH?
- The definition is somewhat unclear.

The film thickness can range as follows (for metals):

- Minimum (10 monolayers of moisture)
- Dew ($\approx$ 1,000 monolayers of moisture)
- Raindrops ($\approx$ 10,000 monolayers of moisture)

When does condensation occur? At 100% relative humidity or when the surface temperature is below the dew point temperature, in the presence of cracks and delaminations, or when the material is hygroscopic? Condensation mimics the behavior of bulk water in that a large decrease in resistance occurs. For example, 5 monolayers have conductivity two orders of magnitude lower than bulk water, while 20 monolayers have a conductivity one order of magnitude lower. Also, the dissolution of additional surface contaminants can result in a rapid decrease in time to failure: days to seconds. In addition, there are changes in migration behavior, in that larger distances can occur even over alternative surfaces (e.g. conformal coating).

Cracks in components or the PCB can greatly reduce the humidity necessary to induce condensation. This is why ECM in cracked chip capacitors happens relatively rapidly, why conformal coating must adhere to the board surface, and why popcorning results in elevated leakage currents in plastic encapsulated microcircuits (PEMs).

So, from a DfR perspective, do not underestimate RH in controlled environments (telecom, server, storage): e.g. macro-environments are controlled to 40–60% RH (the room), and micro-environments can easily exceed 80% RH (circuit board surface) or greater for long periods of time.

Similarly, do not overestimate the potential for condensation in uncontrolled environments. For example, a temperature increase of 5 C is often enough to prevent condensation. Constant power-on can also do this, while excessive condensation can initiate corrosion even over a conformal coating.

Cleanliness becomes the critical focus. But the housing design (preventing physical water from coming into contact with board surface) also plays a major part in control.

Contamination presents two concerns: hygroscopic contaminants and ionizable contaminants that are soluble in water (e.g. acids, salts). Ionic contaminants of greatest concern are primarily anions: especially halides (chlorides and bromides), because they are chemically aggressive due to their chemical structure. These contaminants are very common in electronics manufacturing processes. They decrease pH; few metal ions found in dendrites are soluble at mid to high pH. Cu dendrites require a pH less than 5 to form; silver ions are soluble at higher pH, which is why silver forms dendrites so easily. Cations primarily assist in identifying the source of anions. For example, Cl with K suggests KCl (salt from human sweat). There are multiple sources of these contaminants in the PCB fabrication process. Examples include the rinse water, fluxes, handling, and the storage environment.

Bromide sources are associated with surface processes, e.g. soldermask, marking inks, and fluxes. IPC-TR-476A describes bromide as follows: "Bromide in epoxy resin can diffuse to the surface during a high temperature process such as soldering" (even more so with LF!).

Fluxes are chemicals used for preparing metal surfaces for soldering. They have high-molecular-weight chemistries and are slightly acidic.

For optimum behavior, you want maximum activity during reflow and minimum activity after reflow, which can be a difficult balancing act. Fluxes are separated into the following nomenclatures:

- Rosin only (RO)
- Rosin, mildly activated (RMA)
- Rosin activated (RA)
- Water soluble
- Low residue (no-clean)

More than 75% of the market uses no-clean, 20% uses water soluble, and the remainder use rosin materials. No-clean fluxes use activators (weak organic acids, 0.5 to 5%); wetting agents (surfactants, 1%); and polyglycols (polyethylene glycol and polypropylene glycol), alkylenic ethoxylate, and fluorinated esters. What are the residues of no-clean soldering?

- Water-soluble dicarboxylic acids
- Hygroscopic polyethylene glycol ethers
- Weak organic acids (WOAs): benzoic, butyric, formic, lactic, malonic, oxalic, propionic, succinic, citric, glutaric, adipic, malic

With optimum flux implemented during the reflow process, acids are neutralized after soldering and residual wetting agents are minimized. However, there are issues that need attention during a DfR analysis. Solder beading and solder balling must be controlled. There is the potential for poor adhesion of conformal coatings due to the residue, an issue could arise due to an incompatibility with wave or selective soldering fluxes, and there may be an issue with probing during ICT testing.

Where else can contamination come from?

- Salts from human contact (KCl and NaCl)
- Cleaning chemicals in storage areas
- Outgassing from other materials in the manufacturing area
- Polymeric materials
- The use environment (dust, evaporated sea water, and industrial pollutants)

Voltage must also be considered when examining issues with contaminants. Voltage is the primary driver for electrodissolution and ion migration. For electrodissolution to occur, the applied voltage must be

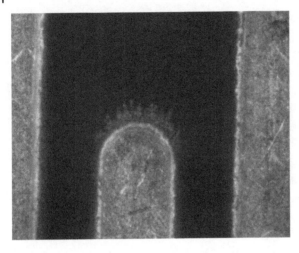

Figure 3.37
E-Field and dendritic growth. Source: Ansys-DfR Solutions.

greater than the EMF: for example, 0.13 V for tin-lead, 0.25 V for nickel, 0.34 V for copper, 0.8 V for silver, and 1.5 V for gold. The velocity of the ions is a function of the electric field strength with regard to ion migration. As the industry has continued reducing the size of electronic packaging, the volts/mil parameter indicating susceptibility has gone up. This increases the risk of an ECM-related failure.

Figure 3.37 shows an example of dendritic growth on ImAg plating in an applied environment: 85 C/85% RF at 10 volts. As can be seen, the migration is only at the tip of the comb pattern. The dendrites stopped growing due to the maximum electric field being impinged.

DfR Solutions performed two studies: NaCl seeding and conformal coating over no-clean flux residues. Over 300 coupons were tested (IPC-B-25). The spacings evaluated were 6.25, 12.5, and 25 mil, and the test conditions were 40 C at 93% RH, 65 C at 88% RH, and 85 C at 85% RH. No ECM occurred at 25 mil spacings. As such, care must be taken in today's designs, as 25 mil spacings are rare.

Conductors must also be examined as part of a review. These include:

- SnPb (solder)
- Sn alloy (LF solder, lead plating)
- Copper (traces, connectors, component leads)
- Silver (conductive adhesives, thick film resistor)
- Nickel (capacitor electrodes, lead plating)
- Gold (connector plating)

Silver will migrate under temperature/humidity/and bias unless special precautions are taken. This might include alloying with an anodically stable metal such as palladium or platinum, using an organic coating, or using a metallic coating (on SMT resistors). Gold requires contamination to migrate (typically chloride).

Numerous time-to-failure models can be applied to these chemical reactions. For fundamental chemical behavior, use the Arrhenius or Eyring equation. For experimental observations, use Barton-Bockris or DiGiacomo. And for similar corrosion-based mechanisms, use Peck's Law or the Howard or CAF equation.

All this activity leads to the cleanliness requirements that must be addressed as part of a complete DfR analysis.

3.10.10 Cleanliness

Based on prior experience with field issues of electronics with contamination, a set of recommended levels has been established for some anionic species on PCB assemblies, as presented in Table 3.6. These levels are especially applicable to uniformly applied contamination.

PCB fabrication can leave process residues that tend to be spread relatively evenly across surfaces. IPC's original contamination specification of 10 microgram/in^2 equivalent NaCl applied to bare PCB manufacturing. Assembly processes introduce flux residues from soldering. A key

Table 3.6 Contaminant cleanliness limits.

Contaminant	Upper control limit (mg/in2)	Maximum level (mg/in2)
Bromide (Br$^-$)	10	15
Chloride (Cl$^-$)	2	4
Fluoride (F$^-$)	1	2
Nitrate (NO3$^-$)	4	6
Nitrite (NO2$^-$)	4	6
Phosphate (PO4$^-$)	4	6
Sulfate (SO4$^-$)	4	6
Total weak organic acids	50	100

difference between this type of contamination and PCB fabrication contamination is localization. Flux residues can be concentrated around solder joints or under components. Because the IPC method is to extract contaminants from an entire PCB or PCBA and calls for the extracted contamination to be divided by the entire surface area of the board, the levels of flux residues measured may be much lower than actual levels in concentrated areas.

Contamination-related failures have been noted on boards where the full board had weak organic acid levels measured below recommended levels. In these cases, visible flux residues are common, typically around solder joints and under components. Interpretation of these recommended levels should account for the possibility of localized residues.

Excessive contamination on PCBAs is often associated with failures from decreased resistance between conductors that are designed to be electrically insulated. The decrease in resistance and consequent increase in leakage current can be due to ECM, CAF, or a conductive electrolyte spanning the conductors. As the spacing between conductors decreases with advancing technology, the levels of contamination that result in failures are also decreasing. This has led to tighter controls on cleanliness.

Let's look at the common contaminants that must be properly cleaned.

3.10.10.1 Chloride

Sources of chloride include an aqueous wash process using water that has not been fully deionized. Reliance on the cleanliness of municipal water can result in intermittent contamination issues because water treatment plants may not tightly control levels of some ions. This is especially true in countries other than the US. Another source of chloride is handling with bare skin, in particular in warm, humid environments where sweat on the skin is more likely. Residues of some flux chemistries that are not halide-free may also contain chloride.

3.10.10.2 Bromide

For epoxy-glass laminate, surface bromide levels typically fall within the range of 0–7 microgram/in2, depending on the amount of flame retardant added by the panel and prepreg manufacturer. Exposure to reflow conditions tends to increase the porosity of the laminate and soldermask and increases the level of bromide extracted from the sample. With several exposures to reflow conditions, bromide can reach levels as high as 10–12 microgram/in2. Bromide levels less than 10 microgram/in2 are not typically considered detrimental on organic PCBs. However, levels between 10 and 15 microgram/in2 may increase the risk for failures if attributable to corrosive flux residues. Levels above 15 microgram/in2 should be considered a significant risk of failures, especially if attributable to corrosive flux residues.

3.10.10.3 Cations

By themselves, cations are not considered a reliability risk unless present in relatively high quantities. In addition, they do not participate in the same chemistries that anions do. Therefore, Ansys-DfR does not have recommended levels. Cations are sometimes useful in determining sources of the more detrimental anions. Calcium and magnesium are often detected when aqueous wash has not been properly deionized. Sodium, potassium, and ammonium are often used as counter ions to active anion ingredients in cleaning chemistries, various processing chemistries, and fluxes. Sodium may also be introduced by handling with bare skin.

3.10.10.4 Weak Organic Acids

WOA levels from flux can vary greatly. They are commonly found in liquid wave flux and solder paste fluxes. Levels can vary depending on the liquid delivery method (e.g. foam or spray) and, for solder pastes, the preheat and reflow dynamics. Low-solids (no-clean) fluxes typically use WOAs as their primary active ingredients. The amount of detected WOAs is often proportional to the amount of residual flux. Bare PCBs typically do not have WOA residues, but HASL processes typically use strong fluxes that must be thoroughly cleaned afterward and may be detectable if not properly removed.

3.10.10.5 Cleanliness Testing

The PCB industry has been interested in the ionic cleanliness of PCB and PCBA surfaces and its correlation with corrosion, ECM, dendritic growth, and subsequent opens, leakage current, or shorting during testing and in the field. Initial methods for cleanliness evaluation included resistivity of solvent extract (ROSE). These techniques were often performed during the manufacturing process, either inline or in batches offline, and measured the electrical conductivity or resistivity of the wash or rinse liquid. Many are still used today as process indicators in PCB manufacturing. One disadvantage of these methods is their inability to detect specific ionic species generating conductivity.

Ion chromatography has more recently become an important technique for evaluating ionic cleanliness. This technique, which detects individual types of ions, allows quicker troubleshooting of contamination sources and better predictions about the detrimental effects of individual ions. It is a form of high-pressure liquid chromatography (HPLC) that uses ion-exchange columns to separate ionic species over time. The separation power is based on several factors including the ionization equilibrium, or pKa, of each ion. A conductivity detector, or in some applications a UV or fluorescence detector, is used for detection as each species leaves the column. Columns and systems are customized and optimized for specific applications. Each system is tailored to detect specific ions or a specific group of ions.

ANSYS-DfR uses three separate columns and systems to detect inorganic anions, cations, and WOAs. The cations and anions systems have a detection limit of ranging from 10 to 100 ppb (10–100microgram/L), depending on the specific ion. WOAs are more difficult to detect because of their lower conductivity values and therefore have a higher detection limit range of 100 to 200 ppb (100–200 microgram/L) depending on the specific acid anion.

The maximum levels of contaminants recommended are listed in Table 3.6. The table shows both the upper control limit and maximum levels under which you should have no reliability issues.

Another potential reliability issue is whether non-functional pads should be used in the fabrication of the circuit board.

3.11 Non-Functional Pads

There is an ongoing debate regarding the influence of non-functional pads (NFPs) on printed board (PB) reliability, especially as related to barrel fatigue on plated through vias with high aspect ratios. A survey of PCB fabricators was conducted to ascertain which way the industry was leaning. The overwhelming response indicated that most suppliers remove unused/non-functional pads. No adverse reliability information

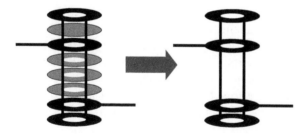

Figure 3.38 Non-functional pads example. Source: G. Caswell and C. Tulkoff. Non-functional pads: Should they stay or should they go? SMTA ICSR, 2014.

was noted with respect to the removal of unused pads; conversely, leaving them can lead to an issue called *telegraphing*. In all responses, whether removing or keeping NFPs, the primary reason given was to improve the respective fabricators' processes and yields. Companies that remove the unused pads do so primarily to extend drill-bit life and produce better vias in the boards, which they considered the primary reliability issue. For those that keep the unused pads, the primary reason given is that they believe it helps manage Z-axis expansion of the board due to CTE stresses (Caswell and Tulkoff 2014). Figure 3.38 illustrates the difference between NFP configurations.

Knowing your PCB fabricator's practice can be critical to your DfR assessment.

3.12 Wearout Mechanisms

This section identifies several ways that wearout can impact today's electronics.

3.12.1 IC Wearout

Because complex ICs within designs may face wearout or even failure within the period of useful life, you need to investigate the effects of use and environmental conditions on these components. The main concern is that submicron process technologies drive device wearout into the regions of useful life well before wearout was initially anticipated to occur.

A DfR analysis for IC wearout uses component accelerated test data and PoF-based die-level failure mechanism models to calculate the failure rate of IC components during their useful lifetime. IC complexity and transistor behavior are contributing factors during the calculation of the failure rate. Four failure mechanisms are modeled using readily available, published models from the semiconductor reliability community, NASA/JPL, and research from Dr. Joey Bernstein from Bar-Ilan University. These mechanisms are:

1. Electromigration (EM)
2. Time-dependent dielectric breakdown (TDDB)
3. HCI
4. Negative bias temperature instability (NBTI)

Taking the reliability bathtub curve (Figure 3.39) into consideration, research shows that EM and TDDB are considered steady-state failure modes (constant random failure), whereas HCI and NBTI have wearout behavior.

Each of these failure mechanisms is driven by a combination of temperature, voltage, current, and frequency. Traditional reliability predictions assume temperature and (sometimes) voltage as the only

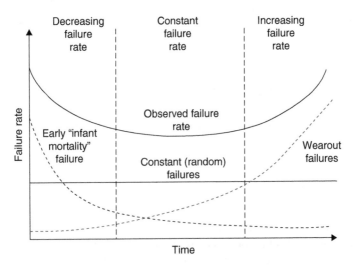

Figure 3.39 Typical bathtub curve.

accelerators of failure. Each failure mechanism affects the on-die circuitry in a unique way. Therefore, each must be modeled independently and later combined. This DfR approach uses circuit complexity, test and field operating conditions, derating values, and transistor behavior as mathematical parameters. Since there is no single dominant parameter set, each mechanism could make the largest contribution to a component's failure rate. In general, there is no dominant failure mechanism; thus, for a specific component, as few as one or as many as all four mechanisms can affect it.

A brief examination of the four failure mechanisms is prudent (DfR Solutions 2011).

EM affects the interconnects in an IC. It is characterized by the migration of metal atoms in a conductor downstream of the electron flow. The result of EM is opens or voids in some portions of the conductor and corresponding hillocks in other portions. It is worth noting that although EM affects metal (wires and vias), it is a prevalent failure mechanism for semiconductor devices because stress migration and atomic diffusion of metals occur at the metal portion of metal oxide semiconductor (MOS) devices and the conductors between them.

TDDB is a failure mechanism that design rules cannot prevent. Design rules do, however, limit the total gate dielectric area exposed to a given electric field. The gate dielectric breaks down over a long period of time due to a comparatively low electric field, relatively speaking, between technology nodes. In advanced nodes, and considering dimensions of features, larger fields are worse. The breakdown is caused by the formation of a conducting path through the gate oxide to the substrate due to electron tunneling current. If the tunneling current is sufficient, it causes permanent damage to the oxide and surrounding material. This damage results in performance degradation and eventual failure of the device. If the tunneling current remains very low, it increases the field necessary for the gate to turn on and impedes its functionality. Research has shown that in smaller technology nodes (starting around 0.1 micron), TDDB has become time-independent or has a constant failure rate. In larger nodes, several breakdown sites will progressively weaken the dielectric; but in smaller nodes, the area of the breakdown site makes up a larger and larger percentage of the total area of the oxide – whereas one breakdown site can cause immediate failure.

HCI occurs in both conducting nMOS and pMOS devices stressed with drain-bias voltage. High electric fields energize the carriers (electrons or holes), which are injected into the gate oxide region. The degraded gate dielectric can then more readily trap electrons or holes, causing a change in threshold voltage, which in turn results in a shift in the subthreshold leakage current. HCI is accelerated by an increase in bias voltage and is predominately worse at lower stress temperatures, e.g. at room temperature. Therefore, HCI damage is unlike the other failure mechanisms as its damage is not replicated in a high-temperature operating life (HTOL) test, which is commonly used for accelerated testing (typical HTOL conditions are to operate the device under test at a maximum ambient temperature calculated from the power dissipation and junction temperature of the device while at maximum operating voltage ratings).

NBTI occurs only in pMOS devices stressed with negative gate bias voltage at elevated temperatures. Like HCI, degradation occurs in the gate oxide region, allowing electrons and holes to become trapped. NBTI is driven by smaller electric fields than HCI, which makes it a more significant threat at smaller technology nodes where smaller electric fields are used in conjunction with smaller gate lengths. The interface trap density generated by NBTI has an inverse proportionality to oxide thickness.

Due to these failure risks, the ICs in your design must be examined for IC wearout.

3.13 Conformal Coating and Potting

Conformal coatings are used on PCBs with the intent of providing protection from harsh environments containing moisture and contamination, such as dust and metallic debris, that could cause shorts in electronic components. In addition, some conformal coatings are designed to provide thermal insulation, shock-vibration attenuation, and electrical insulation for high-voltage components at high altitudes (Qi et al. 2009). Conformal coatings are also used to help mitigate the risk of tin whiskers on pure tin surface finishes (Woodrow and Ledbury

Table 3.7 Conformal coating selection.

	Properties	Comments
Epoxy	Good adhesion	Difficult to rework
	Excellent chemical resistance	Needs compliant buffer
	Acceptable moisture barrier	Not widely used
Urethane	Good adhesion	Difficult to rework
	High chemical resistance	Widely used
	Acceptable moisture barrier	Low cost
Acrylic	Acceptable adhesion	Easy to rework
	Poor chemical resistance	Widely used
	High moisture resistance	Moderate cost
Silicone	Poor adhesion	Possibility of rework
	Low chemical resistance	Moderate usage
	Excellent moisture resistance	High cost
Paralyene	Excellent adhesion	Difficult to rework
	Excellent chemical resistance	Moderate usage
	Excellent moisture resistance	Extremely high cost

2005). These attributes make conformal coatings especially attractive for high-reliability application environments.

There are several types of conformal coating, as shown in Table 3.7. Let's look at them more carefully.

3.13.1 Silicone

Silicone conformal coatings are most widely used in high-temperature environments due to their innate ability to withstand prolonged exposure to higher temperatures than most other conformal coating chemistries. This attribute has made them the primary choice for underhood automotive applications. They can also be applied in thicker films, making them useful as a vibration dampening/isolation tool if the coated assembly is to be placed in a high-vibration environment. Rework of silicone-coated assemblies can sometimes be difficult due to their chemical resistance and the fact that, unlike acrylics and polyurethanes, they do not vaporize with the application of heat.

3.13.2 Polyurethane

Polyurethane formulations provide excellent humidity resistance and far greater chemical resistance than acrylic coatings. They require very lengthy cure cycles to achieve full or optimum cure. Removal of polyurethane coatings can be difficult due to their very high resistance to solvents. For single components, the preferred method of removal is via burn-through for spot repair, while the use of specially formulated strippers enables the user to completely remove the coating from an entire PCB assembly for more wide-ranging rework concerns.

3.13.3 Epoxy

Epoxy coatings are very hard, usually opaque, and good at resisting the effects of moisture and solvents. Epoxy is usually available as a two-part thermosetting mixture and shrinks during curing, leaving a hard, difficult-to-repair film. It possesses excellent chemical and abrasion resistance but can cause stress on components during thermal extremes. Epoxy is quite easy to apply but nearly impossible to remove without damaging the components.

3.13.4 Acrylic

Acrylic conformal coatings are perhaps the most popular due to their ease of application and removal and forgiving nature. Acrylics dry rapidly, reaching optimum physical properties in minutes; are fungus resistant; and provide long pot life. Additionally, acrylics give off little or no heat during cure, eliminating potential damage to heat-sensitive components. They do not shrink during cure, have good humidity resistance, and exhibit low glass transition temperatures. The material also has a continuous operation range of −65 C to +125 C.

3.13.5 Superhydrophobics

A new class of nano-thin coatings that provide superhydrophobic capabilities have entered the marketplace with great fanfare. *Super-hydrophobicity* is defined as having a wetting angle greater than 90 degrees, and some materials have measured wetting angles exceeding 150 degrees. Hydrophobicity tends to be driven by the number and length of the fluorocarbon groups and the concentration of these groups on the surface. The key points of each technology are as follows:

- Some are chemical vapor deposition (CVD) processes, with low vacuum requirements.
- Some use a room-temperature deposition process.
- A variety of potential coating materials are available (with a primary focus on fluorocarbons).
- Some incorporate nanoparticles into a conventional conformal coating.

The applied thickness of these materials is significantly thinner than conventional materials (100 nm or less). They have optical transparency, RF transparency, and rework ability and can eliminate the masking process. These technologies are likely the conformal coating processes of the future.

Potential concerns regarding solder joint integrity arise when the conformal coating material is allowed to flow underneath the package. Recent investigations have shown that letting conformal coating flow underneath plastic quad-flat no-lead package (PQFN) dramatically reduces the thermo-mechanical fatigue (TMF) life of the device by changing the equivalent plastic strain of solder joints from predominantly shear to an axial loading mode (Vianco and Neilsen 2015). In such a condition, the conformal coating expands and contracts in the vertical direction. As thermal cycling progresses, failure in solder joints could result from lifting of the component that causes excessive tensile and compressive stresses. The amount of axial stress greatly depends on the leverage of the conformal coating on the component and the variation of CTE and elastic modulus (E) of the coating with temperature. In addition, conformal coating materials with a Tg that is within the thermal cycle range can drastically reduce the fatigue life of solder interconnects.

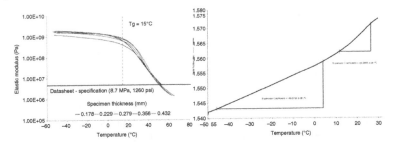

Figure 3.40 Conformal coating Tg behavior.

As the temperature approaches the material's Tg, significant expansion occurs, along with a reduction in material stiffness. If conformal coating material expansion occurs prior to adequate softening, large stresses are applied to solder joints. This behavior is inherent in thermoset polymers since materials expansion tends to be driven by changes in the free volume, while changes in the modulus tend to be driven by increases in the movement of the polymer chains. Figure 3.40 illustrates this behavior.

Potting and conformal coatings share similar attributes and, therefore, issues. Table 3.8 provides some definitions that are used in this section of the chapter.

Potting should provide thermal improvements and environmental protection and not adversely impact the reliability of the driver. The main issue with potting is high expansion, which can generate large deformations and pressures in the circuit card. Potting materials are designed to protect electronics from environmental, chemical, mechanical, thermal, and electrical conditions that could damage the product. The wrong potting selection can cause damage due to unwanted stresses or heat. Some of the criteria for selecting a potting compound are as follows:

- Does the potting compound perform a thermal function?
- Does it need to protect from aggressive chemicals or moisture?
- Does it need to protect from shock loads?
- Will the potting see high temperatures during manufacturing?
- Are issues such as outgassing, cryogenic operation, and medical compatibility involved?

Finding a material that matches all these criteria is very difficult, and trade-offs are necessary. Typically, the cost increases as specific properties

Table 3.8 Potting definitions.

Term	Definition
Coefficient of thermal expansion (CTE)	The fractional change in size per degree change in temperature at a constant pressure
Dynamic mechanical analysis (DMA)	Used to locate the glass transition temperature based on mechanical changes (stiffness) of the material; also used to determine the modulus.
Modulus of elasticity	Mathematical descriptor of a material's elastic deformation when a force is applied to it. An applied force generates a deflection. When the force is removed, the material resumes its original shape.
Shear modulus	Modulus of rigidity (G) is the ratio of shear stress to the shear strain. It can be mathematically related to the modulus of elasticity by Poisson's ratio.
Shear stress	The component of stress coplanar with a material cross-section (in plane loading).
TMA testing	Thermo-mechanical analysis. Measures a change of a dimension of a material due to changes in temperature.
Adhesion	Overlap shear testing to measure the strength of the bond between materials per ASTM D-1002-72.
Cure shrinkage	Variations in the resin volumetric changes during the cure.
Viscosity	
Moisture absorption	Capacity of a polymer to absorb moisture from its environment.
Glass transition temperature (Tg)	The temperature below which the physical properties of plastics change to those of a glassy or crystalline state.

are required. The properties of the potting depend on the material used, formulation, and fillers. Most pottings have a very high CTE (greater than 100 ppm/C), which is the amount they expand when heated. These CTEs are typically greater than those of the housing and electronics (between 5 and 24 ppm/C). This mismatch causes large displacements, which can stress the electronics. Therefore, the hardness or stiffness of the potting is critical (more critical than the expansion). The stresses in the electronics are not linear with respect to temperature. This is due to the glass transition temperature of the potting material. The glass transition temperature defines the temperatures at which the material properties of the polymer change from rubbery to crystalline. Below the glass transition temperature, the material is crystalline and, therefore, much harder. Typically, the CTE decreases, but not to a degree that reduces the stress on the electronics. The Tg is a very important material property that must be considered when selecting a potting material. Epoxies are typically used below their Tg, and silicones are used above their Tg. Other materials, such as polyurethanes, may have multiple Tg or be used around the Tg temperature, making the stress states nonlinear.

Many different materials are used for potting; the two most common are silicones and epoxies. Other materials, such as polyurethanes and asphalt, are also used. Their attributes are as follows:

- Acrylic potting compounds are not commonly used for bulk potting of electronics. Acrylics are UV- and heat-hardening materials, which make them unsuitable for very thick applications such as potting. When used for conformal coating, acrylics are very fast-hardening, have good chemical resistance, and are clear in appearance.
- Asphalt materials have been accepted as potting compounds and coil impregnates for close to 100 years. They are typically used in less complex electronics like transformers and simple power supplies. Asphalt materials have some limitations and may not be suitable for certain applications.
- Epoxy resins are common potting materials and have been used for many years. Epoxies are generally hard, but there are softer formulations. Epoxies exhibit low shrinkage on cure and can be exothermic during the cure. Therefore, it is important to control the temperature during curing to prevent damage to the electronics. Epoxies have very good mechanical properties and temperature performance and good

adhesion. Epoxies used for pottings may have glass transition temperatures in the use range and high coefficients of thermal expansion.

- Polybutadiene urethanes are very resistant to water attack. Polybutadiene is the second-most-produced polymer in the world and, therefore, one of the cheapest polyurethane potting compounds available. The ease of variation of process characteristics and final properties is leading to its increasing use in electronics and electrical encapsulation.
- Polyurethane resins are typically rubbery in their cured states, but there are some much harder formulations. Polyurethanes have lower heat generation during cure than epoxies. Therefore, the heat generated during cure is not usually a problem. Very hard polyurethanes (Shore D) are known to cause failures of power supplies, while soft polyurethanes (Shore A) provide less stress.
- Silicone is one of the most common potting materials used. However, it is also one of the more expensive. Two types of silicone are used for potting: silicone gel and silicone rubber. Silicone gel is very expensive, and the volumes required to fill a cavity often make it cost-prohibitive. Another issue with silicone materials is their susceptibility to sulfur penetration. If sulfur is in the environment, it has the potential to penetrate the silicone and cause dendrites, which may result in shorting. So, again, knowing your environment is critical.

Understanding the properties of your conformal coating or potting material is a vital element of a DfR assessment.

This chapter has offered a myriad of topics that are critical to the performance of a DfR assessment of your product.

References

Bayle, F. and Mettas, A. (2010). Temperature acceleration models in reliability predictions: Justification & improvements. In: Proceedings of the Annual Reliability and Maintainability Symposium (RAMS), 1-6. San Jose: CA. doi: 10.1109/RAMS.2010.5448028.

Benson, R., Phillips, T., Bargeron, C. et al. (1993). Electromigration of silver in low-moisture hybrids. In: Proceedings of the SPIE, 530–536.

Breach, C. (2010). What is the future of bonding wire? Will copper entirely replace gold? *Gold Bulletin* 43 (3): 150–168.

Caswell, G. (2017). Surviving the heat wave - a presentation on thermally induced failures and reliability risks created by advancements in electronics technologies. IMAPS Symposium.

Caswell, G. and Binfield, S. (2017). Contamination and cleanliness: Developing practical responses to a challenging problem. DfR Solutions Conference.

Caswell, G. and Keener, M. (2016). How mitigation techniques affect reliability results for bgas. IMAPS Device Packaging Conference.

Caswell, G. and Tulkoff, C. (2014). Non-functional pads: Should they stay or should they go? SMTA ICSR.

DfR Solutions. (2011). Integrated circuit wearout. Sherlock validation document.

DiGiacomo, G. (1982). Metal migration (Ag, Cu, Pb) in encapsulated modules and time to fail model as a function of the environment and package properties. In: Proceedings of the IRPS, 27–33.

Hallberg, O. and Peck, D. (1991). Recent humidity accelerations, a base for testing standards. *Quality and Reliability Engineering International* 7 (3): 169–188.

Hillman, C. (2013). No MTBF? Do you know MTBF? DfR Solutions white paper.

Hillman, C., Kittlesen, G., and Schueller, R. (2014). A new (better) approach to tin whisker mitigation. DfR Solutions white paper.

Ireson, W. and Coombs, C. (1989). *Handbook of Reliability Engineering and Management.* New York: McGraw Hill.

Kong, R. (2013). Upgrading the component derating process. DfR Solutions white paper.

Krumbein, S. (1988). Metallic electromigration phenomena. *IEEE Transactions on CHMT* 11 (1):5–15.

Lall, P., Pecht, M., and Hakim, E. (1997). *Influence of Temperature on Microelectronics and System Reliability: A Physics of Failure Approach.* Boca Raton, FL: CRC Press.

Levine, L. and Deley, N. (2004). Update on high volume copper ball bonding. www.processsolutionsconsulting.com/pdf/CU/IMAPS04CU .pdf.

Marderosian, A.D. and Murphy, C. (1977). Humidity threshold variations for dendrite growth on hybrid substrates. 15th International Reliability Physics Symposium, Las Vegas, NV. doi: 10.1109/IRPS.1977.362777.

NI. (2017). Redundant system basic concepts. https://www.ni.com/en-us/innovations/white-papers/08/redundant-system-basic-concepts.html.

Qi, H., Osterman, M., and Pecht, M. (Jan. 2009). Design of experiments for board-level solder joint reliability of PBGA package under various manufacturing and multiple environmental loading conditions. *IEEE Transactions on Electronics Packaging Manufacturing* 32 (1): 32–40. doi: 10.1109/TEPM.2008.2005905.

Rudra, B. and Jennings, D. (1995). Electrochemical migration in multichip modules. *Circuit World* 22 (1): 67–70.

Schenkelberg, F. (2018). Norris-landsberg solder joint fatigue. Accendo Reliability. https://accendoreliability.com/norris-landzberg-solder-joint-fatigue.

Sharon, G., Caswell, G., and Blattau, N. (2015). Predicting package level failure modes in multi-layered packages. IMAPS Symposium.

Tulkoff, C., Caswell, G., and Hillman, C. (2013). Best practices for improving the PCB supply chain II: Performing the process audit. SMTAI.

Vianco, P. and Neilsen, M.K. (2015). Thermal mechanical fatigue of a 56 I/O plastic quad flat nolead (PQFN) package. In: *SMTA International Conference Proceedings*, 85–94.

Woodrow, T. and Ledbury, E. (2005). Evaluation of conformal coatings as a tin whisker mitigation strategy. IPC/JEDEC 8th International Conference on Lead-Free Electronic Components and Assemblies.

Zamanzadeh, M., Meilink, S.L., Warren, G.W. et al. (1990). Electrochemical examination of dendritic growth on electronic devices in HCL electrolytes. *Corrosion* 46 (8): 665–671.

4

Design for the Use Environment: Reliability Testing and Test Plan Development

4.1 Introduction

Product test plans are critical to the success of a new product or technology. They need to be stressful enough to identify defects yet show a correlation to a realistic environment. The recommended approach is to employ both industry standards and reliability physics analysis (RPA). This approach results in an optimized test plan that is acceptable to management and customers (Caswell 2013).

Using reliability physics to facilitate the design, performance, and resulting interpretation of accelerated life tests, starting at the design stage of a product and continuing throughout the life cycle of the product, is the ideal methodology for creating a viable test plan.

It is useful to start with industry specifications, modify them as necessary, and then tailor the test strategy specifically for the product, set of materials, use environment, and reliability metrics required. However, industry testing frequently falls short due to a limited degree of mechanism-appropriate testing. Often, mechanism-specific coupons are used that do not reflect the real product, and the test data is hidden from end-users.

Similarly, JEDEC tests are often promoted to original equipment manufacturers (OEMs). These tests, typically for 1000 hours, hide wearout behavior as the use of a simple activation energy along with the incorrect

Design for Excellence in Electronics Manufacturing, First Edition.
Cheryl Tulkoff and Greg Caswell.
© 2021 John Wiley & Sons Ltd. Published 2021 by John Wiley & Sons Ltd.

assumption that all mechanisms are thermally activated, and can result in an overestimation of failures in time (FITs) by 100× or more.

Creating a test plan that uses reliability physics as its basis is the most cost-effective method for determining a product's ability to perform over its intended lifetime. Again, *reliability physics* is defined as the use of science (physics, chemistry, etc.) to capture an understanding of failure mechanisms and evaluate useful life under actual operating conditions. Armed with this knowledge, an engineer can evaluate design and performance, and correctly interpret the results of accelerated life tests, starting at the design stage and continuing throughout the life cycle of the product.

Failure of a physical device or structure (i.e. hardware) can be attributed to the gradual or rapid degradation of the material(s) in the device in response to the stress or combination of stresses the device is exposed to. Common stresses are thermal, electrical, chemical, moisture, vibration, shock, mechanical loads, etc. Failures may occur prematurely, gradually, or erratically.

A viable test plan has defined test objectives that may include test by comparison, qualification testing, validation testing, research, compliance testing, regulatory testing, and failure analysis. Each of these has a specific goal in mind for the test approach, so having a clear understanding of the appropriate objective is a critical first step.

4.1.1 Elements of a Testing Program

Using the objective, now define the elements of the testing program. Start with the creation of a reliability goal for the product, which can be defined in several ways:

- The *desired lifetime* can be defined as the time when the customer is satisfied with the product. This can be long after the warranty has expired. For example, users might be upset if a failure occurs in the electronics of their refrigerator, automobile, television, etc. even 5 to 10 years after the warranty period. As such, the test-plan approach should be actively used to properly define the development of the product and its qualification.
- Determine the lifetime of a product based on the number of returns during a warranty period.

- Define the survivability of a product over a set time with a set confidence level.
- Use a calculated mean time between failures (MTBF) or mean time to failure (MTTF). The recommendation is to avoid this actuarial approach as it is essentially obsolete and no longer very accurate. MTBF and MTTF calculations tend to assume that failures are random, which does not motivate failure avoidance. The calculations are easy to manipulate, and tweaks such as modifying the quality factor for each component are often used to reach the desired MTBF. This approach is also often misinterpreted, as a 50,000-hour MTBF does not mean there are no failures in 50,000 hours. Using MTBF to calculate the lifetime of a product is more suited to logistics and procurement than to failure avoidance.

Typical life expectancies for different products vary over a wide range. The following list provides insight into these lifetimes.

- Low-end consumer products (toys, etc.) Less than 1 year?
- Cell phones: 18 to 36 months
- Laptop computers: 24 to 36 months
- Desktop computers: 24 to 60 months
- Medical (external): 5 to 10 years
- Medical (internal): 7 years
- High-end servers: 7 to 10 years
- Industrial controls: 7 to 15 years
- Appliances: 7 to 15 years
- Automotive: 10 to 15 years (warranty)
- Avionics (civil): 10 to 20 years
- Avionics (military): 10 to 30 years
- Telecommunications: 10 to 30 years
- Solar: 25 years (warranty)

Knowing the desired lifetime is important, as electronics have changed over the past couple of decades. There is now the potential for device wearout, which did not previously exist. Figure 4.1 illustrates the change in the typical bathtub curve as a result of these electronics developments.

When dealing with life expectancies of these lengths, it is necessary to also understand the wearout failure mechanisms that impact a product's ability to meet these them. Wearout mechanisms can be broken down into several categories:

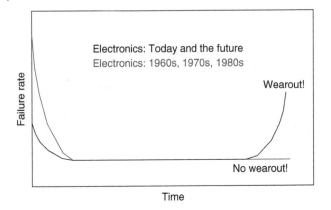

Figure 4.1   Changes to the typical bathtub curve.

Chemical or contaminant:
- Moisture penetration
- Electrochemical-migration driven dendritic growth
- Conductive anodic filament (CAF)
- Corrosion
- Radiation damage

Mechanical:
- Fatigue
- Creep
- Wear

Electrical:
- Electro-migration driven (molecular diffusion and inter-diffusion)
- Thermal degradation

When overstress issues are detected:
- Verify suppliers are meeting material strength specifications and purity expectations.
- Re-evaluate field loading and stress expectation used to design the part.
- Sort out stresses
  As combined stresses are often involved, it is necessary to re-evaluate the effectiveness of product durability testing.

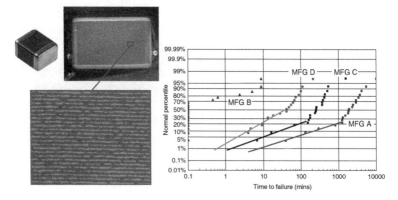

Figure 4.2　MLCC life expectancy. Source: Ansys-DfR Solutions.

In Chapter 5, there is an illustration that depicts the IC wearout mechanism and how that failure mechanism has evolved due to the continued reduction in gate geometries for transistors. The data shows that products with features of less than 90 nm are at risk of not being able to survive 10 years in the field.

Figure 4.2 illustrates the end-of-life mechanism for multilayer ceramic capacitors (MLCCs) where increased capacitance levels coupled with lower voltage levels create a high capacitance-to-voltage ratio that can result in capacitor failure in less than one year at rated voltage and temperature. The variation between suppliers is also an issue. The same capacitor (with regard to max voltage and temperature levels) from different suppliers is not guaranteed to behave the same.

Solder wearout also must be considered when developing an appropriate test plan for a product. As technology has moved from leaded packages like quad flat packs (QFPs) to ball grid arrays (BGAs) to quad flat pack no leads (QFN) to flip chip, the number of cycles to failure has dramatically decreased. QFPs could survive more than 10,000 cycles; BGAs 3,000 to 8,000 cycles; QFNs 1,000 to 3,000 cycles; and flip chip less than 1,000 cycles. This transition to lower-cycle-count failure was principally driven by the reduced compliance in the solder joints. Increases in operational frequencies and higher data-transfer rates have also impacted solder fatigue. Finally, designs today are denser and use lower voltage levels but higher currents that create joule heating. All these trends have resulted in less robust package designs.

Beyond device packaging, consider other components and materials that are susceptible to wearout. The following list delineates these parts and materials along with the issue that drives the failure:

- Ceramic capacitors: dielectric breakdown
- Electrolytic capacitors: electrolyte evaporation, dielectric dissolution
- Resistors: improper derating
- Silver-based platings: exposure to corrosive environments
- Relays and other electromechanical components: wearout models not well developed
- Connectors: improperly specified and designed
- Tin whiskers
- Integrated circuits: next-generation feature size
- Interconnects: creep, fatigue, plated through-holes (PTHs), solder joints

Now that we have defined our reliability goals, we can address the design of the product to ensure that the tests chosen do not create scenarios where the tests drive failures that would not be encountered in the field. This activity involves analyzing the bill of materials (BOM) to ascertain the characteristics of each component, particularly the minimum and maximum operating temperatures for each component.

4.1.2 Know the Environment

The critical first step is having a good understanding of the shipping and use environment for the product. Figure 4.3 illustrates the difference between an ambient environment and the thermal rise associated with the container. The almost 30 C difference has a strong impact on viable test duration to demonstrate life expectancy for a product.

Do you understand the customer and how they use your product (even the corner cases)? How well is the product protected during shipping (truck, ship, plane, parachute, storage, etc.)? Do you have data, or are you guessing? What are the temperature and humidity levels? Is thermal cycling involved in the environment? What is the ambient temperature or operating temperature range? Are salt, sulfur, dust, fluids, etc. inherent to the environment? If so, special testing may be required to ensure compliance with these environments. And finally, are mechanical cycles (lid cycling, connector cycling, torsion, etc.) involved?

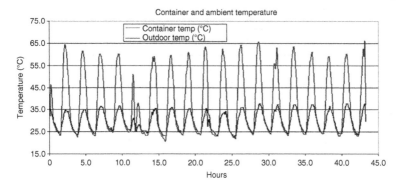

Figure 4.3 Temperature variation in a trucking container.

Many companies respond to questions about the environment with the specification range that the product must operate over: for example, -40 to 85 C. This range is not a diurnal stress range but is more typically a test range. As such, properly defining the field environment is critical.

Table 4.1 illustrates the minimum and maximum temperatures encountered in Death Valley, California, on an annual basis. The average diurnal delta T is approximately 16 C. For comparison, in Quebec, Canada, the diurnal temperature range is from about -7 to 25 C over the year, but the delta T per month is approximately 12 C. The delta temperature is close to the same anywhere in the world, but the average minimum and maximum temperatures are uniquely different based on location. All these parameters are critical in the process of developing a viable test plan for a product.

As a best practice, use standards for structuring the test plan when aspects of the environment are common or when there is no access to information about the use environment. Measure your field environment when some aspects are unique or there is a strong relationship with the customer and this data can be directly gathered. Do not mistake test environments for the actual use environment. This is common for both vibration loads and thermal-cycling ranges.

Let's look at an environment definition example:

Product: Flow monitor on oil pipelines
Environment description

- Outside in the housing (exposed to direct sunlight)

Table 4.1 Climate in Death Valley, CA.

	Jan	Feb	Mar	Apr	May	Jun
Average high in 0 C	19.4	22.9	27.8	32.5	38.1	43.3
Average low in 0 C	4.4	7.9	12.7	16.7	22.6	27.3
Avg precipitation in mm	10	13	8	3	1	1
	Jul	Aug	Sep	Oct	Nov	Dec
Average high in 0 C	46.9	45.9	41.4	33.8	25.1	18.4
Average low in 0 C	31.1	29.8	24.2	16.4	8.9	3.5
Avg precipitation in mm	2	3	5	2	5	8

- On continuously (low power dissipation)
- Fluid flow on pipelines creates vibration
- Shock during shipping and installation

To develop an optimized approach for the testing:

- Temperature: Use the Phoenix environment
- Thermal: Imaging outside of the housing; thermocouple on hottest components in the housing
- Vibration: Measurements with friendly customers
- Shock: SEC 60068-2-27

4.2 Standards and Measurements

A good starting point for a test plan is to use industry or military specifications such as those listed here:

- MIL-STD-810
- MIL-HDBK-310
- SAE J1211
- IPC-SM-785
- Telcordia GR3108
- IEC 60721-3, etc.

The advantage of using these specifications is that they have consensus among the industry, are typically comprehensive, and don't drive additional cost. The issues with these specifications are that they are over 20 years old in many cases and always produce a test result that is less or greater than actual data by an unknown amount. Also, some tests promoted to OEMs are of limited duration, which hides wearout behavior. And the use of an activation energy, with the incorrect assumption that all mechanisms are thermally activated, can result in overestimation of FIT by 100× or more. So, you must take care.

A second approach is based on actual measurements of similar products in similar environments to determine the average and worst-case scenarios. This includes all use environments from manufacturing, transportation, storage, and into the field. Understanding the loads that a product may experience is the next activity.

4.3 Failure-Inducing Stressors

Numerous stressors can impact a product and help to define the correct set of testing parameters. The following list delineates these stressors.

- Temperature cycling: Tmax, Tmin, dwell, ramp times
- Sustained temperature: T and exposure time
- Humidity: Controlled, condensation
- Corrosion: Salt, corrosive gases (Cl_2, etc.)
- Power cycling: Duty cycles, power dissipation
- Electrical loads: Voltage, current, current density, static and transient
- Electrical noise
- Mechanical bending (static and cyclic): Board-level strain
- Random vibration: PSD, exposure time, kurtosis
- Harmonic vibration: G and frequency
- Mechanical shock: G, waveform, number of events

Figure 4.4 illustrates how these different stressors can affect a product.

4.4 Common Test Types

Let's look at the common tests in more detail.

4.4.1 Temperature Cycling

A good starting point for understanding the impact of temperature cycling is IPC-SM-785. Thermomechanical expansion and contraction

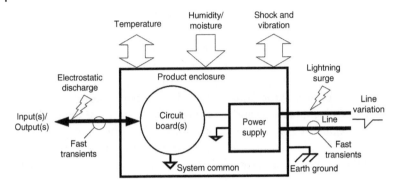

Figure 4.4 Failure load conditions.

are the forces that drive material damage accumulation or stress aging. The primary aging factors are:

- High-end temperature
- High-to-low temperature difference
- Number of cycles
- Correlation to number of cycles (not the time duration)

Secondary aging factors include the dwell time at the hottest extreme and the change rate for the transition from one temperature extreme to the other. Finally, it is useful to limit the extremes for the high-end temperature, the transition rate, and the minimum dwell time at the hot temperature based on the parts mix in the BOM The maximum temperature for the test also *must not exceed* the glass transition temperature (Tg) of the substrate/PWB, as the material properties dramatically change above the Tg. The lowest recrystallization temperature of the plastics used in the device must also be considered, to avoid a test that causes damage that would not be experienced in the field.

Dwell times are also an important parameter when setting up a temperature cycling test. The hot dwell is more important than the cold dwell. Hot dwell is needed to induce creep damage. The hot dwell under a tensile load causes faster attachment aging rates than under a compressive load. Note that for an FR4 PCB, tensile loading occurs at hot temperatures. A practical minimum temperature is essentially determined when cooling parts below 50% of the absolute temperature melting point for the product. Metal becomes a structure (does not creep) at less than 50%

absolute (K) melting temperature. Following through on the calculation, eutectic solder melts at 183 C or 456 K, so 50% = 228 K or -44 C.

The components most susceptible to this type of stress are chip components, crystals, oscillators, filters, and area array devices like BGAs, CSPs, and chips on board (COBs). The issue is due to the coefficient of thermal expansion (CTE) mismatch between the component and the circuit board. Ceramic components (e.g. resistors, alumina substrates) are also susceptible, as are devices having the more brittle Alloy-42 leads. Leaded devices with a copper lead frame do not have this issue.

In a similar vein, long-term constant temperature testing is not recommended. The failure mechanisms tend to be well known and designed for (time-dependent dielectric breakdown [TDDB], electromigration [EM], electrolyte evaporation, etc.), and the acceleration factor tends to be relatively low (long test time to qualify field performance).

4.4.2 Temperature-Humidity-Bias Testing

There are some typical rules of thumb regarding a temperature-humidity-bias environment. A typical test is non-condensing, which would be the standard during operation, even in outdoor applications due to power dissipation. However, a condensing environment can occur in sleep mode or when the system is non-powered. A condensing environment can also occur if driven by the mounting configuration (attached to something at a lower temperature) or if driven by a rapid change in the environment, which can lead to standing water if condensation is on the housing. Finally, a standing-water situation can occur if indirect spray, dripping water, submersion, etc. are used for the test.

If a condensing environment is anticipated, then cyclic humidity testing might be more relevant. This is where repeated applications of condensation events lead to wearout-type behavior (Beta greater than 1) over time as initial condensation events weaken the circuit by inducing dissolution of conductor material. Relevant approaches are defined in MIL-STD-810, IEC 60068-2-30, and IPC TM-650 2.6.3.1 / 2.6.3.4.

In a non-condensing situation, the best practice is to perform a step-stress approach using 40 C / 93% relative humidity (RH) for 72 hours at bias followed by 65 C / 88% RH for 72 hours at bias. It has been noted that the typical testing with 85 C and 85% RH had issues. A study by Sohm and Ray (Bell Labs) demonstrated the degradation of weak organic acid residues above 55 C, which reduces their effect on surface insulation resistance. Turbini demonstrated the breakdown of polyglycols at elevated temperatures as well. Finally, absorption into the PCB can increase the risk of CAF. Regardless of the test condition, continuous functional monitoring is the best practice. The user should maximize bias but minimize power dissipation.

4.4.3 Electrical Connection

From a reliability perspective, the connection to the electrical power source should not be an issue, as the grid is well-defined in developed countries. However, with globalization, electronics are being used increasingly in third-world countries. This is where issues can arise. Some rules of thumb: China can have issues with grounding in facilities due to poor grounding rods. India has numerous brownouts and has extremely high transients on the power bus. Similarly, Mexico has high-voltage surges. When structuring a test plan, these conditions must be considered to determine the viability of your product for these markets.

4.4.4 Corrosion Tests

Salt Spray or Fog

Although there are 14 different industry testing protocols for salt spray/fog/mist the most highly implemented is ASTM B117 which allows the test duration to be structured in 24 hour increments based on the application. For high reliability applications follow the direction herein. This test should be performed per ASTM B117, "Standard Practice for Operating Salt Spray (Fog)," using the following test conditions: 35 C chamber temperature, salt fog solution per specification, 96-hour exposure, and non-operating electronics.

The samples should be positioned horizontally, which is more severe than vertically since the settling of the fog is heavier on horizontal surfaces. Examine the product after exposure, clean gently with a soft brush to remove the salt without affecting the assembly, and then power up and test.

Corrosive Gases

The mixed flowing gas (MFG) test is used to simulate corrosive mechanisms found in equipment exposed to industrial environments. It consists of Cl_2, H_2S, SO_2, and NO^x gases and is most commonly used for

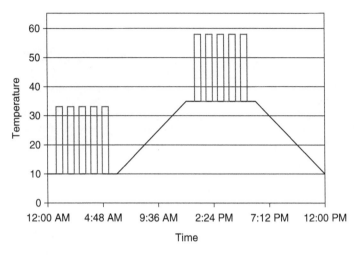

Figure 4.5 Power cycling.

connectors or products in particularly corrosive environments (in-home disposal systems, paper processing, etc.).

4.4.5 Power Cycling

Cycling the power in addition to standard thermal cycling is a way to perform qualification reliability testing where an acceleration factor greater than 120 can be attained. Acceleration is achieved through an increase in frequency. Figure 4.5 illustrates this concept.

4.4.6 Electrical Loads

These stresses involve voltage, current, and current density being applied to the product to more closely simulate the actual operating environment. These stresses can either be static loads or transient loads.

4.4.7 Mechanical Bending

The primary model for this type of test is a single overstress event. This type of event occurs primarily during manufacturing and installation with depanelization, in-circuit testing (ICT), heatsinks, and connector-insertion processes are known to be the most likely causes for the stresses. Figure 4.6 illustrates these operations.

Depending on the location, magnitude, and speed with which the flexure occurs, components and solder joints on the board can be

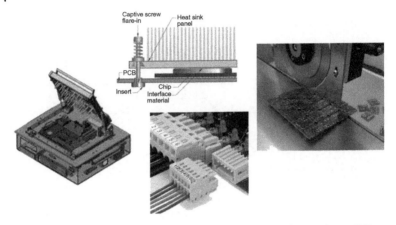

Figure 4.6 Manufacturing operations impacting bending. Source: Ansys-DfR Solutions.

damaged. Common failure modes include ceramic capacitor flex cracking, as shown in Figure 4.7 (left) and pad cratering (right).

Flexure or bending of a PCB is influenced by the load applied, area and location of the load, area and location of support, PCB material, and PCB thickness. Component or interconnect failure is then determined by board-level strain.

The tests for this are addressed in two industry specifications: IPC/JEDEC 9702, "Monotonic Bend Characterization of Board-Level Interconnects"; and IPC/JEDEC 9704, "Printed Wiring Board Strain Gage Test Guideline." Microstrain levels below 500 microstrain are recommended, as few failures occur at that level. At 1000 microstrain, serious issues have been observed, particularly with BGAs and ceramic capacitors.

Similarly, cyclic bending is driven by push-button applications (e.g. cell phones) and is addressed by JEDEC JESD22B113, "Board Level Cyclic Bend Test Method for Interconnect Reliability Characterization of Components for Handheld Electronic Products."

4.4.8 Random and Sinusoidal Vibration

During exposure to vibration, the circuit card assembly responds by cyclic deflection in a manner that corresponds to its natural frequencies. These

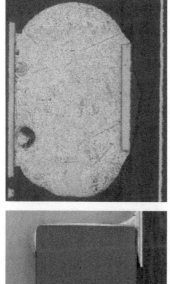
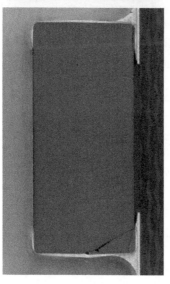

Figure 4.7 Cracked capacitor and pad cratering. Source: Ansys-DfR Solutions.

natural frequencies are dependent on the geometry, stiffness, mass, and boundary conditions of the circuit card assembly.

Exposure to vibration loads can result in highly variable results. Vibration loads can vary by orders of magnitude (e.g. 0.001 G^2/Hz to 1 G^2/Hz), and time to failure is very sensitive to these loads. There is a very broad range of vibration environments, with the MIL-STD-810 alone listing 3 manufacturing categories, 8 transportation categories, 12 operational categories, and 2 supplemental categories. Table 4.2 shows these environments.

Vibration loads can be very complex. They can be sinusoidal (g as a function of frequency), random (G^2/Hz as a function of frequency), and sine(over/on random). Vibration loads can also be multi-axis and can be damped or amplified, depending on the chassis or housing. As such, the response of the electronics is dependent on attachments and stiffeners. Peak loads can also occur over a range of frequencies with a typical range of 20 to 2000 Hz (Hillman 2017).

Failures primarily occur when peak loads occur at frequencies similar to the natural frequency of the product/design.

Natural frequency examples:

- Larger boards, simply supported: 60–150 Hz
- Smaller boards, wedge locked: 200–500 Hz
- Gold wire bonds: 2–4 kHz
- Aluminum wire bonds: greater than 10 kHz

Measuring vibration loads is principally done using accelerometers to measure the acceleration of the printed wiring board (PWB) during vibration. The acceleration can be used to calculate board displacement and curvature. Strain gauges can also be used to directly measure the board-level strain, but they require a more complex setup for performance. Figure 4.8 is an example of a strain gauge that has been bonded to the board surface.

The failure of solder joints due to vibration is based on a technique like the one developed by Steinberg (2000). The main modification is converting from displacement-based criteria to board-level strain criteria. The Steinberg equation for critical PCB deflection at which a component will survive 10 million cycles during harmonic vibration or 20 million cycles during random vibration is:

$$Z_{3sigmalimit} = 0.00022B/Chr\sqrt{L}$$

Table 4.2 MIL-STD-810 vibration environments.

Section 2	Description
2.1	**Manufacture/Maintenance**
2.1.1	Category 1 - maintenance/manufacturing processes
2.1.2	Category 2 - shipping and handling
2.1.3	Category 3 - environmental stress screen (ESS)
2.2	**Transportation**
2.2.1	Category 4 - truck/trailer/tracked - restrained cargo
2.2.2	Category 5 - truck/trailer/tracked - loose cargo
2.2.3	Category 6 - truck/trailer/tracked - large assembly cargo
2.2.4	Category 7 - Aircraft-jet
2.2.5	Category 8 - Aircraft - propeller
2.2.6	Category 9 - Aircraft - helicopter
2.2.7	Category 10 - Ship - surface ship
2.2.8	Category 11 - Railroad - train
2.3	**Operational service**
2.3.1	Category 12 - Fixed-wing aircraft - jet aircraft
2.3.2	Category 13 - Fixed-wing aircraft - propeller aircraft
2.3.3	Category 14 - Rotary wing - helicopter
2.3.4	Category 15 - Aircraft stores - assembled jet aircraft
2.3.5	Category 16 - Aircraft stores - material - jet aircraft
2.3.6	Category 17 - Aircraft stores - assembled, material, propeller aircraft
2.3.7	Category 18 - Aircraft Stores - assembled, material, helicopter
2.3.8	Category 19 - Missiles - tactical missiles - (free flight)
2.3.9	Category 20 - Ground vehicles - ground mobile
2.3.10	Category 21 - Watercraft - marine vehicles
2.3.11	Category 22 - Engines - turbine engines
2.3.12	Category 23 - Personnel - material carried by/on personnel
2.4	**Supplemental considerations**
2.4.1	Category 24 - All material - minimum integrity test
2.4.2	Category 25 - All vehicles - cantilevered external material

Figure 4.8 Strain gauge. Source: Ansys-DfR Solutions.

where:

- B is the length of the PCB parallel to the component
- c is a component packaging constant (typically between 0.75 and 2.25)
- h is the PCB thickness
- r is a relative position factor and is 1.0 when the component is located at the center of the PCB
- L is the component length

The main issues with this formula are that it is limited to simple board geometries (since the maximum deflection is always assumed to be at the center of the PCB) and it doesn't account for board curvature.

There are two types of circuit board vibration durability issues. The first involves the PCB in resonance: components can be shaken off or fatigued by the board motion due to flexing attachment features, as shown in Figure 4.9. The time to failure is determined by the intensity/frequency of stress vs. the strength of the material.

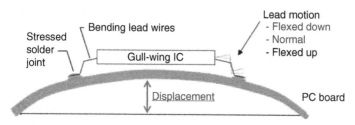

Figure 4.9 Vibration durability issue 1.

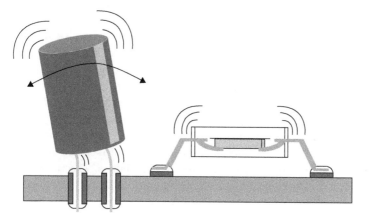

Figure 4.10 Vibration durability issue 2.

The second issue involves components in resonance: components shake and fatigue themselves apart or off the board, as shown in Figure 4.10. This scenario occurs with large, tall, cantilevered devices, aluminum electrolytic capacitors, small leaded sensors, and large coil assemblies.

The following are best practices associated with vibration testing:

- Based on acceleration factor derived from end-of-life simulation. Most common: based on the specification provided by the customer.
- Continuous functionality testing. Most common: periodic functionality test. Avoid functionality tests before and after testing (relatively worthless).

- Test to failure (provides data for continuous improvement). Most common: test to life (based on acceleration factor) or test to spec (based on specifications).

4.4.9 Mechanical Shock

Mechanical shock testing is heavily based on military and industry specifications since random events are difficult to capture and characterize. The primary driver for failure is out-of-plane displacement, like vibration, where solder joint failure is sensitive to intermetallic thickness. Correlation to the field environment is based on the Arrhenius equation using an activation energy of 0.5 eV.

Mechanical shock issues are normally initially encountered during shipping and transportation. This has become a more significant issue due to the increased use and importance of portable electronic devices, particularly those for medical applications. Testing involves the height of a drop for the shock, the surface the device is dropped on, the orientation of the device being dropped, and the number of drops the product might see during its expected lifetime. The test is also used to identify potential embrittlement or damage associated with lead-free (LF) solder that could reduce robustness during exposure to drop conditions. The stiffer LF solder may cause damage to the PWB and create solder intermetallics. The standard failure modes are either bond pad pullout or intermetallic fracture.

Drop testing does not usually require the development of an acceleration factor: e.g. the number of drops in test = number of drops in the field, and the severity of drops in test = severity of drops in field. However, preconditioning is required to replicate aging in the field, as an increase in intermetallic thickness can result in a decrease in solder robustness. Figure 4.11 illustrates how increasing the preconditioning time and temperature results in a decreasing Weibull slope.

4.4.10 ALT Testing

Accelerated life testing (ALT) is always valuable in determining an appropriate test time since it provides an acceleration factor, ensures efficient use of resources, and avoids the possibility of irrelevant failures under test. All ALT tests are designed as test-to-pass. Failures can cause confusion/delay without a high degree of certainty in test relevance.

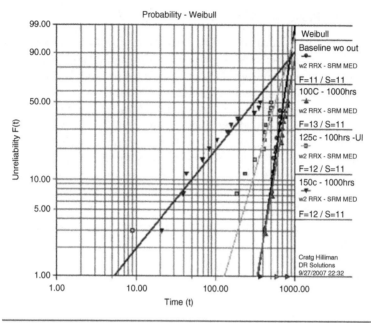

β1=5.8118, η1=766.2208, ρ=0
β2=4.9112, η2=854.5502, ρ=0.9808
β3=2.8634, η3=642.4329, ρ=0
β4=1.0477, η4=429.6169, ρ=0.9727

Figure 4.11 Preconditioning Weibull slope change.

ALT-RGT (reliability growth test) is not a pass/fail test. It is a set of development learning procedures used to identify the operating and failure limit envelope of the unit under test (UUT); and to identify the design margin (i.e. safety factor) of the operating and failure envelope relative to the specified operating ranges through step-stress probe testing of temperature, voltage, and mechanical shock/vibration excitations. It is also used to identify opportunities to improve the reliability and durability of the UUT by expanding the design margins and uncovering hidden inherent design and manufacturing weaknesses in the UUT. Doing so creates effective, permanent corrective actions that can be developed to eliminate field-relevant weaknesses and measure circuit board modal vibration response to determine if it is consistent with long fatigue life criteria.

4.4.11 Highly Accelerated Life Testing (HALT)

In highly accelerated life testing (HALT), a product is introduced to progressively higher stress levels to quickly uncover design weaknesses, thereby increasing the operating margins of the product. This translates to higher reliability. The product is tested in operational mode while the vibration stress is increased with each thermal cycle. The objective of the test is to cause failure of the product. This identifies the weakest link, which can be then be improved. The test duration is typically less than a week. On its own, this test is not able to predict the life of a product since the acceleration factor is not known. However, it is very useful when a product can be compared side-by-side with a previous generation of the product with known reliability.

How does HALT testing compare to the more traditional reliability testing?

HALT:
- Gathers information on product limitations
- Focuses on design weaknesses
- Six degrees of freedom (DoF) vibration
- High thermal rate of change
- Loosely defined; modified on the fly
- Not a pass/fail test
- Results used as the basis for highly accelerated stress screening (HASS) or environmental stress screening (ESS)

Traditional testing:
- Simulates a lifetime of use
- Focuses on finding failures
- Single-axis vibration
- Moderate thermal rate of change
- Narrowly defined; rigidly followed
- Pass/fail test
- Results not typically used in ESS

HALT takes you through combined stresses to technology limits of the product: step stressing (individual and combined), testing with a powered product with monitored tests, and finally, root cause failure analysis and appropriate corrective action.

To properly execute a HALT test, begin with cold step stress and then proceed to hot step stress. Step in 10 C increments, and as you approach the product's "limits," reduce to 5 C increments. Set the dwell time minimum to 10 minutes + time to run functional tests to ensure that the product is still functional. Start dwell once the product reaches the temperature set point. Begin functional tests after a 10-minute dwell. Continue until the fundamental limit of technology is reached. If circuits have thermal safeties, ensure operation, and then defeat the limits to determine the actual operating and destruct limits.

For the vibration step stress, understand the vibration response of the product (i.e. how does the product respond to increases in vibration input?). Determine Grms increments (usually 3 to 5 Grms on the product). Set the dwell time minimum to 10 minutes plus the time to run functional tests to ensure that the product is still operational. Start the dwell once the product reaches the vibration set point. Continue until the fundamental limit of technology is reached.

Post testing involves determining the root cause of all failures that occurred. Once the analysis has been completed, meet with the design engineers to discuss the results of developmental HALT and the root cause analysis. Determine and implement corrective action. Perform a verification HALT to ensure that all problems have been fixed and new problems have not been introduced. Finally, periodically evaluate the product as it is subjected to engineering changes.

4.4.12   EMC Testing Dos and Don'ts

Do leave adequate space on the PCB for electromagnetic interference (EMI) suppression devices. With good EMI design practices, the circuit board may be very quiet. But to be on the safe side, leave extra space on the board for the addition of filters and chokes. It is also a good idea to add extra pads to the board in case a shunt capacitor is needed.

Do not assume that second-sourced parts will have the same spectral characteristics as the primary parts. This is especially important for active components such as DC-DC converters. The fit, form, and function may be the same, but EMI noise could be significant. Before investing in a large stock of these extra parts, evaluate one into the design.

Do use ground and power planes when designing PCBs. And, where possible, use multiple ground planes. Not only are they useful in containing high-frequency traces, but they also help reduce the loop areas of signal and power traces, which are a significant contributor to EMI emissions. In general, when it comes to ground traces, the more copper, the better.

Do use wiring harnesses and wire ties when routing cables inside a box. EMI pickup on I/O cables is a significant contributor to overall radiated noise. Route cables along the sides of the box and away from high-frequency components and switching power supplies.

Do not paint seams on a chassis or box. This is especially important if one of the EMI strategies is containment. Rivets and screws often do not provide adequate electrical conductance between chassis parts. Mask the mating pieces before painting or powder coating. For added suppression, have each mating piece meet with a flange (MET Labs 2014).

4.5 Test Plan Development

Preparing a viable test plan involves several steps to properly identify the requirements for the tests. This subsection identifies a methodology for this test plan approach. It discusses the necessity for reviewing the BOM to determine part limitations, assessing the field environmental conditions so they can be properly mapped to the tests implemented, and reviewing the impact of failure history, should it exist. Next, it is necessary to generate the acceleration factors for the test protocols. Determining the acceleration factors involves identifying the failure mechanisms or environments the unit has encountered. There are several formulas for determining the acceleration factor based on the anticipated failure mechanism, some of which are discussed in Chapter 3. Next, identifying the correct activation energy for the materials being tested is also critical. An energy of 0.7 eV has classically been the activation energy for integrated circuits, but it is not applicable for all device types. Finally, create a test protocol that facilitates accelerated testing. Doing so requires an understanding of the reliability metrics for the product (e.g. reliability requirement, life expectancy, confidence level, sample size). The test plan can then be formulated, in which the specific test conditions and parameters are defined. These involve temperature range, humidity, cycles to failure, power (power cycling), and unusual stresses like dust or salt fog.

This subsection addresses this process flow and provides insight into the approach to take in developing test plans. The objective is to develop a test plan that does not stress the assembly to levels where a failure might not be induced in the field.

The first step in test plan development is to determine the potential failure mechanisms by reviewing the BOM and determining part limitations and the maximum stresses they can handle. It is also necessary to examine the field conditions and assess failure history if it exists. Simultaneously, you also need to understand the environments that the product will endure in the field.

Then, determine which acceleration factor formula to use (see Chapter 3). Next, determine the activation energy of the components and materials used in the product.

The process shown in Figure 4.12 uses the Sherlock solder fatigue calculator to determine a viable acceleration factor. In the Sherlock solder fatigue calculator, enter the information for the part most susceptible to failure. In this case, it is a 2512 chip resistor. The calculation on the left defines a diurnal thermal cycle from 20 to 40 C. The calculation on the right defines a test cycle from -20 to 85 C with 1 hour at each extreme. The diurnal calculation is 135,132 cycles to failure, while the test is 4358 cycles. The acceleration is the diurnal divided by the test cycles = $Af = 31$.

The next step is to identify the reliability metrics for the product:

- Reliability in percentage (e.g. 95%).
- Confidence in test in percentage (e.g. 90%).
- Life expectancy of the product in years.
- Sample size available for test.
- Use the beta and acceleration factors.

It is also necessary to understand the customer's limitations concerning testing, e.g. sample size, test duration, chamber capabilities, and monitoring requirements. The number of samples is often the hardest decision to make. Most companies ($\approx$85%) subject fewer than 10 samples through any accelerated life test. Some companies ($\approx$10%) test a larger number of samples (16–24) to failure to obtain good distribution behavior. A few companies ($\approx$5%) perform a statistical analysis of sample size. Many more

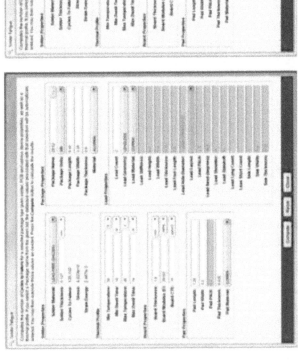

Figure 4.12 Acceleration factor calculations. Source: Ansys-DfR Solutions.

want to but adjust when the calculated sample size is larger than attainable or affordable.

With this information, move into the process.

4.5.1   The Process

Follow these steps to implement the best practice methodology for developing a comprehensive test plan:

- Assemble boards at optimum conditions.
- Rework specified components on some boards.
- Visually inspect and electrically test.
- Perform C-SAM (C-scanning acoustic microscopy) and X-ray inspection on critical components on five or more boards (plus three with rework ed BGAs).
- Use these boards for further reliability testing (TC, HALT, shock, and vibration).
- Perform a failure analysis.
- Compile the results and review.

Once this information has been compiled, start with the definition of the testing plan. Proper failure analysis is also critical to the process. Without identifying the root causes of failure, true corrective action cannot be implemented, and the risk of repeat occurrence increases. Use a systematic approach to failure analysis by proceeding from nondestructive to destructive methods until all root causes are identified. Select techniques based on the failure information specific to the problem, failure history, failure mode, failure site, and failure mechanism.

The first part of this process is to develop a failure history (Tulkoff et al. 2012). When did the failure(s) occur? Were they due to step-stress testing, temperature cycling, vibration, combined, etc.? What stresses were being experienced by the product before failure, e.g. temperature, vibration, combined, including electrical stresses? Was a specific change in the design, BOM, or manufacturing being assessed?

Definitions:
- Failure mode: The failure behavior experienced by the observer
- Failure site: The supposed location of the failure

- Failure mechanism: The mechanism initiating the failure (not the root cause)

The stresses experienced before failure provide a strong indication of potential failure sites. Temperature stress testing tends to drive component issues. High temperatures are often driven by an increase in leakage current (diode turn-on) or an increase in resistance (time delay). Semiconductors, particularly power components, are susceptible. Cold temperatures can also drive failures, primarily in liquid electrolytic capacitors. Cyclic stresses (temperature cycling, vibration) tend to drive interconnect issues, e.g. solder joints, connectors.

4.5.2 Failure Analysis

Performing any type of failure analysis must follow a process that provides maximum information with minimal risk of damaging or destroying physical evidence. This means performing non-destructive evaluation (NDE) techniques first followed by destructive approaches as necessary. Detailed failure analysis technique information is contained in Chapter 5.

Figure 4.13 identifies the various failure modes and how they relate to the different tests.

4.5.3 Screening Tests

The appropriateness of a screening process can be based on three factors: (i) the ability to capture the most common defects, (ii) industry best practices, and (iii) correlation to percent lifetime/time in the field. The effectiveness of a screen is always determined after the fact. It is based on the correlation between defects detected during screening and how they correlate to defects in field failures, when defects are detected (after how many hours or cycles?), and any changes in the early life failure rate.

Screening can capture common hardware defects like microvia pad separation failures in a PCB or optical fiber breaks, as they require very little additional stress to precipitate failure. Sometimes, due to the presence of potential self-healing stresses, failures can be intermittent. The screening then requires exposure to either elevated temperatures or a change in temperature to improve detection capability.

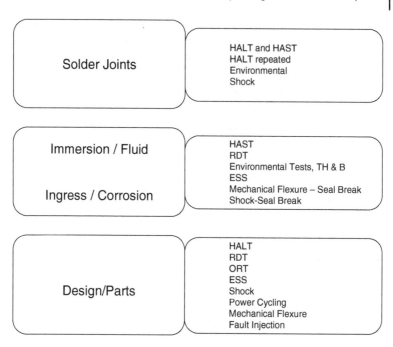

Figure 4.13 Potential failure modes and tests.

In several cases, it has been noted that the failure mode encountered by screening was driven by exposure to the reflow process. Part-level process changes are not always validated through reflow (especially real solder attachment as opposed to simulated reflow). Part-level screening never includes simulated reflow and seldom includes functional thermal cycling (which is necessary to capture failure).

Most enterprise companies tend to focus on detection screens like an ESS, usually involving thermal cycling between the minimum and maximum operating temperatures with full functional verification during thermal cycling. Other screening tests, such as constant temperature, precipitation screens, vibration-shock, and highly accelerated stress screen ing (HASS), are relatively uncommon for ESS. However, the justification for them includes the facts that constant temperature screens are ineffective; precipitation screens are too risky, given system complexity; and mechanical loads are not expected in the field environment (other than transportation).

Studies by the Institute of Environmental Sciences (IES) and the Rome Air Development Center (RADC) have indicated that temperature cycling (ESS) can be twice as effective for detecting defects as a high TEMPtemperature (burn-in).

Typical screening tests range between 2 and 10 cycles for a detection screen. Many companies start at 10 cycles and back off to 2 cycles based on when defects are encountered. Ramp rates vary between 3 and 30 C per minute. Users have reported success in identifying defects when splitting thermal cycles between slow (2 to 5 degrees C/min) and fast (20 to 30 degrees C/min) ramp rates. There seems to be some indication that, due to heat capacity, systems react differently under different ramp rates, creating different stress states (magnitude and location). Dwell times tend to fall into three categories:

- Category 1: Only dwell long enough to test functionality (typically 5–15 minutes).
- Category 2: Dwell long enough to oxidize exposed surfaces (easier to detect intermittent failures; typically 15–60 minutes).
- Category 3: Dwell long enough to perform some burn-in (typically 30–240 minutes).

For detection screens (aka ESS), the temperature range used is almost always the minimum and maximum specified operating limits. A minority of companies will extend the screen to the minimum and maximum measured operating limits – 5 C (based on HALT), which is used to introduce more stress and theoretically capture more defects, but this comes with some risk.

Many organizations believe that how the unit is operated during the detection screen is even more important than operation under ambient temperature. The best practice is to schedule a continuous loop of power-up, functional checks, and power-down with the assurance that each stage (power-up, functional, power-down) occurs at least once during ramp-up, max temperature, ramp-down, and min temperature. Some companies extend dwell times to fit in multiple operational cycles.

Some companies go to more extreme screens depending on the application (primarily military, some aviation) and concerns regarding defects. The most extreme approach is HASS, which combines vibration and thermal cycling and testing out to the measured operating limits.

The purpose of screening is to detect and screen out early life failures driven by the presence of defects. There are two approaches for specifying the definition of *early life*: the first 3–12 months of operation or the first 5% of lifetime. Correlating screening stresses to lifetime can be performed using reliability physics with a combination of simulation software and hand calculations.

4.5.4   Case Study One

Industrial application:

- Environment tends to be high humidity (95% RH).
- Test data suggests the unit will not be exposed to high humidity while running because of the equipment's inherent ability to dehumidify the enclosure.
- Concern is that in some environments, the equipment will operate less frequently and could become soaked with humidity while dormant.
- Environment commonly seaside: salt is an absolute concern.

Typical chemicals used in exposure testing:

- Acetic acid (glacial)
- Acetone
- Ammonium hydroxide (20% by weight)
- ASTM reference fuel C
- Diethyl ether
- Methyl ethyl ketone
- Furfural
- Ethylene dichloride
- Ethyl acetate
- n-Hexane
- Methanol
- 2-Nitropropane
- Toluene

The assembly was also conformally coated with an acrylic resin whose wet thickness was 200–250 microns. After curing, the dry thickness was 50–60 microns. This coating material was designed for use on boards having no-clean flux residue.

The reliability objective for this product was to demonstrate a lifetime of 10 years by running tests with zero failures and a sample size of only six units.

Many potential failure mechanisms can be accelerated on electronics. To handle this variability, the electronics industry has typically assigned an "average" activation energy of 0.7 eV. Based on this activation energy, testing at 105 C provides a 26× acceleration factor over 55 C operating temperature. As such, a 10-year life can be demonstrated by testing for 3309 hours.

Under thermal cycling, we derive the following: for air temperature variation, prior analyses performed has found that 24 cycles of -40 C / 85 C is equivalent to 1 year in a realistic worst-case industrial environment (uncontrolled indoor in Phoenix, AZ). As such, 24 one-hour test cycles equal 365 diurnal cycles, with a test cycle being 20-minute dwells with ramp times of 10 minutes. Using this approach, 48 cycles of -40 C / 85 C = 2 years in the field, as do 64 cycles of -10 C / 85 C or 84 cycles of 30 C / 85 C. This test provides insight into solder joint fatigue and PTH fatigue that result from differences in thermal expansion between the components and the PCB.

Because of the application environment, the potential for dust and water ingress was also assessed. Test standard IEC 60529 was used to test the electronics for resistance to dust and water ingress. In the dust test Method, 13.4 talcum powder is used as the dust medium. Condition 5, category 2 is recommended (no pressure differential between the enclosure and the chamber). Test duration was 8 hours, and a sample size of three was recommended. A thermal coupling test that monitored the temperature rise of the electronics as dust clogged the system was also recommended.

Also, because of the numerous caustic materials noted in the environment, mixed flowing gas (MFG) testing was recommended. The appropriate specification is EIA-364-65, class IIA for 4 days (2 years' equivalent) (336 hours for 10-year life equivalent). Note: this is an expensive test and not a standard test for industrial control equipment.

In the same vein, a salt spray test was also recommended. The standard for this test is MIL-STD-810, method 509.5, or IEC 60512-6. The test has a 96-hour duration, and the electronics typically are not operational during this test. The unit must function after the test is completed.

Finally, to assess the humidity, test methods IEC 61215 10.12 (humidity freeze conditions) and 10.13 (damp heat) were recommended. This test methodology is called out by ASTM E1171-09, which was created to assess the robustness of solar panels. The IEC tests call out 10 humidity

Table 4.3 Product environment conditions.

Environment	Conditions
Operating environment	
Temperature	0 to 50 C
Humidity	5% to 95% non-condensing
Altitude	10,000 feet
Device mechanical strength	IEC-601-1 subclause 21a and b
	IEC-601-1 subclause 21.5
Storage environment	
Temperature	-5 to 55 C
Humidity	5% to 95% non-condensing
Altitude	10,000 feet

freeze cycles (-40 C / 60 C) followed by 1000 hours at 85 C / 85% RH. Power cycling the electronics was also recommended, as it would create a worst-case situation.

4.5.5 Case Study Two

In this case study, the test plan was intended to provide an approach to determine the ability of the connectors in the unit to meet 99% reliability with a 90% confidence level using a small sample size.

The acceleration factor was calculated first. Then, Weibull analysis was used to project test duration:

- Input 99% reliability requirement.
- Input 90% confidence level.
- Input 7-year operational life requirement.
- Input sample size (32 will be required).

The operating environment for the product is shown in Table 4.3.

The Sherlock wearout tool was used to ascertain the acceleration factor for the test, which was from −40 to 85 C. The field temperature (T^0) at 50 C gives a Delta T of approximately 30 C. The Delta Tt (test) is approximately 125 C, tt (time of test) is 1 hour, T^0 = 24 hours (diurnal cycle), Tmax (field) = 50 C, T max (test) = 85 C. The acceleration factor is then calculated to be 12.5 cycles in the field to be equivalent to 1 cycle in test. Therefore, 2555 (assuming 1 thermal cycle per day) field operation cycles is 204 cycles under test. Achieving the 16.15-year duration derived from Weibull indicates that 471 cycles (b = 2.3) would ensure that the connectors could achieve the 99% reliability requirement with a 90% confidence level using 32 samples.

Following the same thinking, raising the maximum test temperature to 125 C, which is the maximum temperature for the connectors, allows a significant reduction in the test time (−40 to +125 C cycle) due to the increase of 40 C in the thermal cycle. This yields an acceleration factor of 31.5, which results in the following: 2555 (assuming one thermal cycle per day) field operation cycles is 82 cycles under test. Achieving the 16.15-year duration derived from Weibull indicates that 190 cycles would ensure that the connectors can achieve the 99% reliability requirement with a 90% confidence level using 32 samples.

The temperature cycling was followed by a seven-day humidity/cycling test using the same samples to ascertain the robustness of the conformal coating or potting. This method does not attempt to duplicate the temperature/humidity environment but instead provides a generally stressful situation that is intended to reveal problem areas with materials – in this case, the conformal coating. A temperature/humidity/bias test for 168 hours at 85 C and 85% RH is recommended. This needs to be done on live units to properly bias them. This test determines whether there are problems with the connectors or any other element of the product.

Only two things can happen to connectors at altitude: they outgas, or they run hotter than at ground level. In this case, it was not felt that an altitude test was necessary. But it was recommended that three units be subjected to a test that simulates 10,000 feet of altitude (equivalent to the inside of a commercial jetliner [8,000 ft]). This requires fully operational units to ascertain if issues from heat rise occur. The test duration should be for one hour.

The vibration requirement called out by the customer appeared to be an adequate test to verify the robustness of the connector interface.

For this equipment and its parts, including mounting accessories, perform a broadband random vibration test under IEC 60068-2-64:2008, using the following conditions (note: this represents Class 7M1 and 7M2, as described in IEC/TR 60721-4-7:2001):

- Acceleration amplitude:
  - 10–100 Hz: 1.0 $(m/s^2)^2/Hz$
  - 100–200 Hz: -3 dB per octave
  - 200–2000 Hz: 0.5 $(m/s^2)^2/Hz$
- Duration: 30 min per perpendicular axis (three total)

Shock testing should also be performed per IEC 60601-1-11. Shock test under IEC 60068-2-27:2008, using the following conditions (Class 7M2 as described in IEC/TR 60721-4-7:2001 using test Type 2):

- Peak acceleration: $300 \, m/s^2$ (30 g)
- Duration: 6 ms
- Pulse shape: Half-sine
- Number of shocks: 3 shocks per direction per axis (18 total)

An insertion-withdrawal test for the contacts would be beneficial due to the number of times the connectors would be separated before replacement. The concern was not with the mating interfaces but rather with fatigue of the surface-mount technology (SMT) solder joints.

A simple insertion-withdrawal test was recommended, where the male and female connectors are removed and reinserted 300 times and then visually inspected to see if any solder joint cracking is occurring. Periodic visual inspections are recommended.

4.5.6   Case Study Three

This customer was replacing an electrolytic capacitor connected directly to line voltage with a solid capacitor mounted behind a surge protection circuit. The use environment was defined as -40 to 65 C for shipping and storage, with a humidity range of 5 to 95%. The ambient temperature around the electronics was -20 to 61 C. The product also saw on/off cycles for a hot region operation of 126,900 total (87,000 cooling; 25,500 heating; 14,400 defrost), cold region operation 154,500 total (34,500 cooling; 90,000 heating; 30,000 defrost). The voltage range was 197 to 253 volts on the input.

The reliability requirement was 94% survival after 60,000 hours of operation and 99.5% during the first 5000 hours of operation. The capacitors had to support continuous operation at 100% of rated voltage and be able to withstand peak voltages up to 315% of rated voltage. They had to support continuous operation from -29 to 68 C, providing the temperature did not exceed 70 C, and maintain continuous operation at all frequencies below 66 Hz from 5 to 95% humidity. Similarly, storage temperature ranges from -40 to 90 C had to be achieved. The components also had to survive handling shock and vibration environments and pass a pass salt spray test per ASTM-B117-73 for 1000 hours.

The aging of metallized film capacitors is fundamentally determined by the aging of the dielectric, electrodes, and end spray contacts. The speed of this aging process is accelerated by temperature, voltage, and humidity. All three influence factors are relevant for film capacitors in industrial applications. This test plan was designed to consider the simultaneous influence factors of temperature, voltage, and humidity.

It is necessary to monitor the capacitance tolerance and equivalent series resistance (ESR) of these capacitors to ascertain failure. Not only can an uncontrolled loss of capacitance have fatal consequences in the application, but it is also important to keep ESR within an acceptable value during the life cycle to limit power loss and self-heating of the film capacitor. An increase in ESR leads to an increase in power loss, as well as corresponding heating of the film capacitor.

The expected lifetime of a metallized film capacitor in operation can be obtained by measurements and calculations, including all relevant factors: temperature (e.g. Arrhenius's law), voltage (exponential law), and humidity (e.g. Hallberg-Peck model).

For the temperature-humidity-bias test, 40 C / 93% RH at a maximum operating voltage (AEC-Q200 test for automotive) and 60 C / 85% RH at a maximum operating voltage (stays below a maximum operating temperature of 70 C) were recommended. These two tests were noted, and higher stress conditions were then considered because, for accelerated life tests to be useful, it is fundamental to complete the test in a realistic time.

In his paper "Comprehensive Model for Humidity Testing Correlation," Stewart Peck (Lau 1992) developed an acceleration formula that provides a direct correlation from autoclave test results up to 140 C to low humidity down to 30% RH. Classically, the standard test for determining the effect of humidity has been the electrically biased test of devices at 85 C with 85% RH. It was felt that a requirement of 10% failures at 1000 hours for the test would indicate reliability in standard operational conditions. His model allows the changing of temperature and humidity to ascertain the acceleration factor for temperature-humidity-bias testing. Peck's equation is as follows:

$$TTF = Ao * RH^{-N} * e^{(Ea/kT)}$$

where:

- TTF= Time to failure
- Ao = Arbitrary scale factor
- RH = Relative humidity (percent)
- N = An experimentally determined constant
- Ea = Activation energy (eV)
- k = Boltzmann's constant
- T = Temperature (Kelvin)

The proposed approach was:

- Test humidity of 95% and field humidity of 70%
- Test temperature of 85 C and field temperature of 35 C
- Activation energy = 0.7 eV
- Humidity power-law coefficient = 2.7

This yielded an acceleration factor of 33.52, which means it would take 60 days to validate a 15-year life if no failures were encountered. This option would mean stressing the parts at their maximum voltage (between 180 and 270 VAC) above the maximum operating temperature, and above the maximum humidity for operation, but it would get results in the fastest time. There is a concern with running a test well above the maximum operating temperature, as the failures encountered may not be indicative of failures in a field environment. Failure analysis would be necessary to ensure that the failure mode corresponds to field history.

The relays needed to be evaluated per the on-off cycles previously noted. Using the endurance curve for the relay, it was determined that the relays could handle 150,000 actuations at maximum current. To complete this test in the same 60 days would require an actuation every 35 seconds.

As can be seen, each type of test plan has unique properties and approaches.

References

Caswell, G. (2013). Test plan development using physics of failure: The DfR Solutions approach. DfR Solutions. https://www.dfrsolutions.com/resources/effective-reliability-test-plan-development-using-physics-of-failure.

Hillman, C. (2017). Test plan development. DfR Solutions. https://www.dfrsolutions.com/hubfs/Resources/services/Test-Plan-Development.pdf?t=1509565547846.

MET Labs. (2014). Top 5 EMC do's and don'ts. www.metlabs.com/pages/emc.html.

Lau, J. (1992). IPC-SM-785: *Guidelines for Accelerated Reliability Testing of Surface Mount Solder Attachments*. Northbrook, IL: IPC.

Steinberg, D. (2000). *Vibration Analysis for Electronic Equipment*. Princeton: Wiley Interscience.

Tulkoff, C., Caswell, G., Blattau, N., and Hillman, C. (2012). Root cause analysis of halt failures. DfR Solutions webinar and white paper.

5

Design for Manufacturability

5.1 Introduction

In the electronics industry, the quality and reliability of any product are highly dependent on the capability of the manufacturer, regardless of whether it is a contract manufacturer or a captured shop. Manufacturing issues are one of the top reasons companies fail to meet warranty expectations, which can result in severe financial pain and eventual loss of market share. What a surprising number of engineers and managers fail to realize is that focusing on processes addresses only part of the issue. Design plays a critical role in the success or failure of manufacturing and assembly.

Design for Manufacturability (DfM) is the process of ensuring that a design can be consistently manufactured by the designated supply chain with a minimum number of defects. DfM requires both an understanding of best practices and an understanding of the limitations of the supply chain. It is the process for optimizing all the manufacturing functions: supplier selection, procurement, receiving, fabrication, assembly, quality, operator training, shipping, delivery, and repair. The goal is to ensure that the critical objectives of cost, quality, reliability, regulatory compliance, safety, time-to-market, and customer satisfaction are known, balanced, monitored, and achieved.

Building on a Design for Reliability (DfR) foundation, successful DFM efforts require the integration of product design and process planning into a cohesive, interactive activity known as *concurrent engineering*.

Design for Excellence in Electronics Manufacturing, First Edition.
Cheryl Tulkoff and Greg Caswell.
© 2021 John Wiley & Sons Ltd. Published 2021 by John Wiley & Sons Ltd.

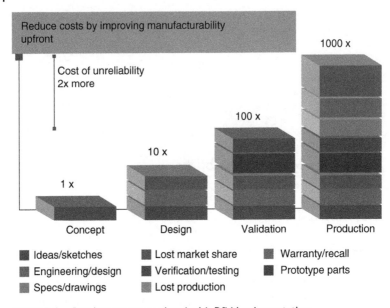

Figure 5.1 Cost increases associated with DfM implementation.

If existing processes are used, new products must be designed to the parameters and limitations of these processes regardless of whether the product is built internally or externally. If new processes are used, then the product and processes need to be developed concurrently and carefully, considering the risks associated with "new." The overall concept is to reduce costs by improving manufacturability, as shown in Figure 5.1.

The foundation of a robust DfM system is a set of design guidelines and tasks to help the product team improve manufacturability, increase quality, reduce life-cycle cost, and enhance long-term reliability. These guidelines must be customized to the user's culture, products, and technologies and must be based on a solid understanding of the intended production system. This chapter contains some global DfM guidelines and tasks that apply to most electronic systems (Tulkoff 2014):

1. Know and understand problems and issues with current and past products related to manufacturability, delivery, quality, repairability and serviceability, regulatory issues, and recalls. Implement strategies to address and prevent recurrence of mistakes, using an effective

system to capture and disseminate this historical knowledge throughout the organization. At an absolute minimum, hold brainstorming sessions after product launch to collect and assess lessons learned and best practices from all areas of the organization.

2. Standardize design, procurement, processes, assembly, and equipment throughout the supply chain. Limit the use of custom components. Doing so reduces overall cycle time, simplifies training and tasks, reduces repeated mistakes, improves the opportunity for bulk discounts, and improves the opportunity for automation and operation standardization. Limit exotic or unique components, as they drive higher prices due to low volumes and less supplier competition. Unique components also result in increased opportunities for supply-chain disruptions.

3. Simplify the design by reducing the number of parts. Parts reduction is one of the best ways to reduce the cost of fabricating and assembling a product and to increase quality and reliability. This activity results in fewer opportunities for defective parts and fewer assembly errors. Develop an approved or preferred parts lists or a standardized bill of materials (BOM), and directly link it back to the computer-aided design (CAD) system. When possible, use one-piece structures from injection molding, extrusions, castings, and powder metals or similar fabrication techniques instead of bolting or gluing multi-part assemblies. Establish part families of similar parts based on proven materials, architectures, and technologies that are scaled for size and functionality. Use parts that can perform more than one function: for example, a cover or base plate that also serves as a heat sink. Incorporate guiding, aligning, or self-fixturing features into housing and structures.

4. Design for lean processes. Lean supply, fabrication, and assembly processes are essential design considerations. Anything that does not add value to a product is waste and should be reduced or eliminated. Lean processes are more likely to be performed quickly and correctly, resulting in reduced throughput time, which equates to faster time to market and lower costs. Designs that are easy to assemble manually are more easily automated; and assembly that is automated is more uniform, more reliable, and of higher quality.

   Wherever possible, develop and use standard guidelines appropriate for the process being performed.

Examples:

- Common hole sizes, lines, and spacings.
- Standard handling: Avoid components with moisture sensitivity levels (MSLs) greater than 3.
- For assembly: Design for human factors (the "visual" factory).
- Allow for visual, audio, and/or tactile feedback to ensure correct assembly operations.
- Provide adequate access clearances for tools and hands.
- Design work to use standard tools and settings: crimpers, splicers, cutters, solder iron tips, drill bit sizes, torque settings, and wire sizes.

5. Design for parts handling by minimizing the handling required to correctly position, orient, and place parts. Avoid multiple or complex assembly orientations. Use non-symmetrical parts where possible. When symmetrical parts are needed, use keying features to ensure proper orientation. Make orienting and mating parts as visually obvious as possible. Use parts oriented in magazines, bands, tape, reels, or strips; or use parts designed to consistently orient themselves when fed into a process. Reduce and avoid parts that can be easily damaged, bent, or broken. Reduce the need for temporary fastening and complex fixtures. Begin assembly with the component that has the largest base and low center of gravity, and add other parts to it.

6. Design for joining and fastening. Fasteners increase the cost of manufacturing, handling, and feeding operations. Avoid threaded fasteners when possible, since they're easy to strip. When fasteners must be used, minimize the variety of shapes and sizes. This also improves serviceability.

7. Use error-proofing techniques. Mistakes happen. Anticipate and eliminate opportunities for error. Create an assembly process that is visually obvious, well-defined, and clear cut, to reduce confusion and the need for interpretation. Minimize wording in work instructions; use pictures, icons, videos, and photographs instead of words. Have revision-controlled written instructions in only one designated location. "Key" unique parts so that they can be inserted only in the correct location and orientation. Design verifiability into the product and its components by using color coding or audio or tactile feedback. Electronic products can be designed to use self-test diagnostics such as built-in self-test (BIST) or Joint Task Action Group (JTAG) functionality.

8. Design for process capabilities. Use specific supplier DfM guidelines, or know the process capabilities of the production equipment expected to be used. Avoid unnecessarily tight tolerances and tolerances that are at the inherent capability of the manufacturing processes or operators. Tighter specifications are not always better and drive additional cost and testing! Perform tolerance stackup analyses on the complete system across the many connected processes and parts. Determine when new production process capabilities are needed, and allow enough time to develop new processes, determine optimal process parameters, and establish necessary process controls.
9. Design for test, repair, and serviceability. Design built-in diagnostics, self-tests, meaningful error messages, and diagnostic interfaces. Minimize disassembly steps and tools required to access replaceable and repairable items. Carefully consider the placement of components like fuses, batteries, fans, and filters that can be serviced in the field.

5.2 Overview of Industry Standard Organizations

Numerous industry standards organizations provide different types of guidance to help facilitate good DfM. While their insights are tried and true, recognize that they represent the minimum acceptable industry requirements. Users need to modify and customize these to fit their products and environments.

ANSI

The American National Standards Institute (ANSI) has created standards with the aim of ensuring the safety and health of consumers and the protection of the environment. The organization has been in existence for 100 years and has produced over 11,500 standards. ANSI primarily links together organizations like the ASTM, IPC, Institute of Electrical and Electronics Engineers (IEEE), and SAE to generate consensus on standards. ANSI's America Makes category works closely with the other organizations on standards for additive manufacturing; and its roadmap shows the work still needed to provide this industry-wide coordination.

IPC

Figure 5.2 illustrates the various IPC specifications that could apply to a DfM assessment.

From a DfM perspective, IPC-2221, "Generic Standard on Printed Board Design," provides general requirements for the design of printed wiring boards (PWBs) and other forms of component mounting or interconnecting structures, whether they're single-sided, double-sided, or multilayer. It provides information on three performance classes for PWBs:

- Class 1, General Electronic Products: Consumer products
- Class 2, Dedicated Service Electronic Products: Communications equipment; sophisticated business machines, instruments, and military equipment where high performance, extended life, and uninterrupted service are desired but not critical.
- Class 3, High-Reliability Electronic Products: Commercial, industrial, and military products where continued performance or performance on demand is critical and where high levels of assurance are required.

Comments on IPC Class 2 vs. Class 3

Great quality is necessary but not *sufficient* to guarantee high reliability, and IPC Class 3 does not guarantee high reliability. A PCB or PCBA can be perfectly built to IPC Class 3 standards and still be completely unreliable in its final application. Consider two different PCB laminates, both built to IPC Class 3 standards. Both laminates are identical in all properties *except* that one laminate has a coefficient of thermal expansion (CTE) of 40 while the other has a CTE of 60. The vias in the laminate with the lower CTE will be *more* reliable in a long-term, aggressive thermal cycling environment than the vias in the CTE 60 laminate. A CTE 40 laminate built to IPC Class 2 could even be *more* reliable than the CTE 60 laminate built to Class 3. Selecting appropriate materials for the end-use environment is key!

JEDEC

JEDEC is the leading developer of standards for the solid-state industry. All JEDEC standards are available online at no charge at www.jedec.org. Some commonly referenced JEDEC/IPC joint standards include:

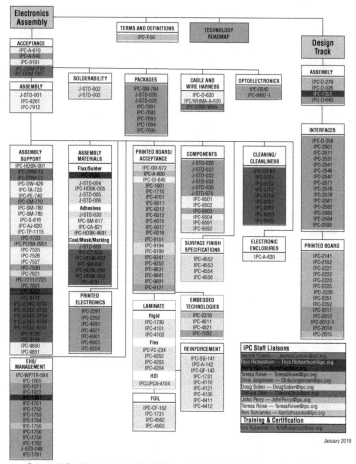

Figure 5.2 IPC Standards 2019. Source: IPC Simplified Standards Tree, IPC International, Inc. © 2019 IPC International, Inc.

- J-STD-020D.01, "Joint IPC/JEDEC Standard for Moisture/Reflow Sensitivity Classification for Non-Hermetic Solid-State Surface Mount Devices": This document identifies the classification level of non-hermetic solid-state surface-mount devices (SMDs) that are sensitive to moisture-induced stress. It is used to determine what classification level should be used for initial reliability qualification.
- JS9704: IPC/JEDEC-9704, "Printed Wiring Board (PWB) Strain Gage Test Guideline": This document describes specific guidelines for strain gauge testing for PWB assemblies. The suggested procedures enable board manufacturers to conduct required strain gauge testing independently, and provides a quantitative method for measuring board flexure and assessing risk levels. The topics covered include test setup and equipment, requirements, strain measurement, and report format.

ASTM

Like ANSI, the American Society for Testing and Materials (ASTM) is a group dedicated to the development of standards using the consensus process. Its focus on materials and mechanical environmental stresses is aligned with a DfM program. The ASTM has created over 12,000 standards covering multiple market segments and industries.

MIL

Military specifications are also an integral element of the testing methods available to use during a DfM assessment. In many cases, military testing approaches are the most viable for assessing harsh environments. The specifications address transportation vibration, mechanical shock, and thermal cycling.

IEC

The International Electrotechnical Commission (IEC) has generated over 6,000 standards that focus on electrotechnology. The standards are developed globally, and each member country gets one vote in determining a consensus. The IEC's testing specifications are used in DfM testing programs and assess many of the typical environments encountered.

IEEE

The IEEE is the world's largest technical professional society and focuses on computing, electronics, and electrical engineering. It consists of 39 separate societies, several of which have links to DfM-related issues. The IEEE Explore library is a major source of support for DfM analyses. The IEEE has over 1,300 standards that focus on the functionality, capabilities, and interoperability of products and services.

UL

Underwriter Laboratories focuses on promoting safe living and working environments through the application of safety science and hazard-based safety engineering. UL addresses the manufacture of products that are physically and environmentally safe, with a specific focus on minimizing loss of life and property. As such, its processes and procedures help companies to demonstrate safety, deliver quality, and provide sustainability through inspection, advisory services, and certification testing services.

These resources are beneficial in performing a viable DfM analysis and should be used to enhance the process.

5.3 Overview of DfM Processes

At this point, it is useful to identify some objectives of a robust DfM process. The importance of understanding DfM can be delineated by the fact that over 70% of manufacturing costs are determined during the design stage; actual production costs drive a much smaller percentage. A viable DfM process further reduces these costs and helps identify manufacturing-related issues even before the initial prototypes are created.

Reducing manufacturing costs is a primary focus of DfM, and the process begins with a streamlined BOM. Simply put, fewer parts mean less purchasing activity, lower inventory, reduced handling and controls on the production floor, simpler inspections, and easier testing protocols. Careful creation and review of the BOM are how this is accomplished. As an example, during a BOM review, it was noted that a customer had eight different 0.1 µF capacitors. The differences included the dielectric chosen, size of the capacitor, and voltage rating. After analysis, it was determined that multiple engineers had designed sections of the schematic and had not compared capacitor selections. By performing this comparison, the eight types were reduced to two. This increased the usage volumes of the two types and significantly reduced the overall cost.

Similarly, when performing BOM analysis, using standard components must be considered. Standard components are significantly less expensive than custom devices. The history and reliability of standard parts are also better understood and characterized. Shorter lead times for standard parts also facilitate on-time delivery of the product, further minimizing costs.

Selecting parts for multi-use is another way to facilitate cost reductions. Although parts may perform different functions in different products, they keep the design criteria simpler and keep costs under control. The key to this activity is to minimize the number of different functions and variations used. If done properly, the result is a standard parts family from which multi-use parts can be selected.

Ease of fabrication should also be considered. The materials selected need to be chosen to help minimize costs. This might entail control of tolerances, availability of raw materials, and ancillary functions that can result in higher production costs.

The capability and accuracy of the pick and place systems for the components selected must also be reviewed. The designer must understand either the internal manufacturing or the contract manufacturer's skillset to ensure that compliance with the BOM part selections can be achieved.

Finally, understanding the handling of components is paramount. Handling analysis could involve the positioning of tape and reeled parts on feeders to minimize pick-up rotations and reduce processing time. Consistent component orientations should be maintained through the production process.

These guidelines are the start of a viable DfM process. The following subsections of this chapter go into more detail on topics critical to this issue.

5.3.1 The DfM Process

In the past, different types of DfM reviews were performed, starting with simple "gut check" reviews performed by engineers relying on their many years of experience. Since this knowledge was not written, this approach had difficulties when companies transitioned products to original design manufacturers (ODMs) in Asia. Other DfM processes used a combination of internal teams and external consultants to perform a formal documented design review. Another approach uses the expertise of larger contract manufacturers (CMs) that do the DfM review as a service. Currently, automated design analysis (ADA) software with modules that automate DfM rule checking is becoming more dominant. Performing a formal DfM review is often overlooked, as the designer lacks the specialized expertise or the design function is completely removed from the manufacturing organization.

DfM reviews need to be performed at the bare board level, at the completed circuit board level, at the point of integrating the chassis and housing, and at the system level. Since many companies use CMs for assembly, it is also necessary to perform the review in conjunction with each CM used. DfM results can be dramatically impacted by different mixes of equipment and processes at different CMs.

Implementing a root cause failure analysis process and problem-solving methodology is also an integral requirement of a robust DfM process. Know the product and process history. Analyze, learn from, and prevent problems, or they will recur.

However, before spending time and money on failure analysis (FA), consider the following:

- Consider analysis order carefully. Some actions limit or eliminate the ability to perform additional tests.
- Understand the limitations and output of the tests selected.
- Use labs that can help select and interpret tests for capabilities you don't have.
- Avoid requesting a specific test. First describe the problem and define the data and output needed.
- Pursue multiple courses of action. There is rarely one test or one root cause that will solve a problem.
- Consider how the data will help solve the problem. Some FA is just not worth doing!

5.4 Component Topics

Numerous failure modes and mechanisms can impact a product. Understanding how they occur and how to prevent them during the design stage can vastly improve a product's ability to withstand the rigors of its intended environment.

Physics of failure (PoF) is a proactive science-based philosophy that addresses material science, physics, and chemistry and provides the basis to develop an up-front understanding of failure modes and mechanisms. Knowing how things fail is equally as important as understanding how and why things work, by enabling engineers and designers to be knowledgeable about the root causes of failures so that they can be designed out in new products. PoF provides a scientific basis for evaluating usage life and hazard risks for new materials, structures, and technologies when exposed to their actual operating conditions.

5.4.1 Part Selection

Parts selection is the process of creating the BOM during the virtual design process before physical layout occurs. The overriding goal should be to keep things simple. New component technology can be very attractive but is not always appropriate for high-reliability systems. Be very conservative. Important actions for part selection include critical component identification and derating (Caswell 2015a, 2016).

5.4.2 Moisture Sensitivity Level (MSL)

Popcorning of components is controlled through MSLs as defined by IPC/JEDEC documents J-STD-020D and J-STD-033B. Identify MSL for all components, and identify the maximum MSL components as well. Not all datasheets clearly list MSL; the information can be buried in reference or quality documents. Ensure that the listed MSL conforms to the latest version of J-STD-020. Components that require extra MSL diligence include aluminum and tantalum polymer capacitors.

During a DfM process, it is vital to understand MSL, as the exposure time on the production floor must be controlled to ensure that moisture in the air does not harm the components in question. Table 5.1 notes the six levels for MSL control and the amount of exposure time permitted before the parts need to be re-baked to drive out moisture.

During a DfM assessment, look at the potential for popcorning in tantalum-polymer capacitors if they are part of the design. Higher lead-free (LF) reflow soldering temperatures lead to increased susceptibility of popcorning of these devices. These parts are typically stored in moisture barrier bags (MBBs), but not all suppliers package the

Table 5.1 MSL levels.

Level	Floor life (duration)	Conditions
1	Unlimited	<300C/85% RH
2	1 year	<300C/60% RH
2a	4 weeks	<300C/60% RH
3	168 hours	<300C/60% RH
4	72 hours	<300C/60% RH
5	48 hours	<300C/60% RH
5a	24 hours	<300C/60% RH
6	Time on label (TOL)	<300 C/60% RH

parts with desiccant to keep the moisture level low. So, as part of the DfM review, confirm the MSL level and perform a more rigorous visual inspection of polymer capacitors during the initial build.

5.4.3 Temperature Sensitivity Level (TSL)

Temperature sensitivity involves limits on maximum process temperatures provided by the component manufacturers. Many component manufacturers rely on J-STD-020 for guidance on this issue. However, there have been issues with specifying a uniform dwell time. J-STD-020C specifies within 5 C of 260 C for 20–40 seconds; while manufacturers desire 260 C for 5 to 10 seconds. Special, lower process temperature specifications are commonly documented for optoelectronic devices. The concern is the potential for latent defects after exposure to the higher temperatures associated with LF reflow temperatures. SnPb is 215 to 220 C peak, whereas LF is 240 to 260 C peak. Initial observations of parts have identified deformed or damaged components that, in many cases, are due to the manufacturer failing to update component specifications. In addition to the optoelectronic devices previously mentioned, electrolytic capacitors, ceramic chip capacitors, and surface-mount connectors are also sensitive to maximum process temperatures and should be monitored during the DfM assessment.

5.4.4 ESD

Electrostatic discharge (ESD)- or electrical overstress (EOS)-related failures can have a significant impact on a product's life expectancy and reliability. ESD is a sudden flow of electricity between conductive and non-conductive surfaces that builds an electric charge and then discharges it to ground, impacting any susceptible electronics in the path (Caswell 2015b).

There are three predominant ESD models at the device level: the human body model (HBM); the charged device model (CDM), and the machine model (MM). There are others for the system level, as defined in IEC 61000-4-2.

The HBM simulates the ESD event when a person charged to either a positive or negative potential touches an integrated circuit (IC) that is at another potential. The CDM simulates the ESD event wherein a device charges to a specific potential and then comes into contact with a conductive surface at a different potential. The MM simulates the ESD event that occurs when a part of a piece of equipment or tool comes into contact with a device at a different potential. HBM and CDM are repeatable real-world models, but MM is deprecated.

Every major IC technology advance has mostly negatively impacted ESD performance. As such, today's ICs may no longer provide adequate ESD for portable electronics. Compounding this, new circuit features and RF applications are increasing the complexity of both ESD and system-level IEC protection.

Two primary failure mechanisms are associated with ESD: silicon dioxide breakdown and thermal destruction. Silicon dioxide breakdown can short the gate of the affected device permanently; or the fields can push carriers into insulators, causing degraded performance. Thermal destruction can occur on any resistance path that is subject to localized intense heating for the static discharge. Contacts, vias, and junctions can be affected, with the weakest link failing first. ESD damage can also produce "walking wounded" – a situation where the devices experience increased leakage and resistance levels as well as softened junctions. This latent damage may continue to accumulate without any additional external failure signature.

The user needs to understand that ESD protection is necessary at the IC, component package, and system level. Multiple approaches are needed to achieve reliable protection. Designing for ESD impacts both product design and manufacturing process controls. Know the ESD rating for each part, and select parts (where possible) for the best ESD rating. Identify all ESD sensitive parts on drawings. Mark locations of ESD sensitive parts on the PCB with the ESD symbol. Design ESD protection for internal ESD sensitive parts to meet ANSI 20.20 Class 2 (2000V). If parts are rated as less than Class 2, then additional protection circuitry will be needed to protect the assembly during handling. Consider the entire system as ESD sensitive, and use ESD protection on all susceptible parts (not just system I/Os). Susceptibility is greater as gate geometries for ICs continue to get smaller.

As part of a DfM analysis, there are several options to determine whether adequate protection has been designed into the product. For example, bandpass filters or polymer ESD devices can be used for external system inputs. For ESD-sensitive internal pins, single-bypass capacitors may suffice, but filters or polymer electro-static discharge devices (PESDs) can be added. Any pin of an ESD-sensitive part may be at risk if it is not connected to a power supply plane, decoupled adequately to ground, or protected by a filter network. External outputs should be protected by low-capacitance clamping diodes or PESDs if speed is an issue.

5.4.5 Derating

Derating is intended to reduce the failure rate of components in a product and improve that product's reliability. Components will fail: the issue is when and under what circumstances. By providing a design standard that identifies derating requirements by product application category, the cost of excessive derating, the cost of inadequate derating, and the impact on reliability can all be avoided.

Derating – or, more properly, design margin planning – is also used to add robustness and reduce failures due to excessive variation in tolerances, power surges, etc. The *derating factor* is defined as the ratio of the component's maximum measured stress divided by the component's absolute maximum rating. Maximum temperatures and temperature rises are defined in Celsius. For derating or design margin planning to work, the components used must be of a known quality level.

Derating requirements apply to worst-case conditions found during normal operation within the voltage, temperature, and load extremes required by the product specification and the category defined. A test process must not overstress any component or material such that its future reliability is impaired. Components used solely for protection do not usually require derating. Exceptions are components with a significant terminal voltage, even if they are generally in an off state. Normal derating for the voltage applied to the component terminals must be applied.

Worst-case derating is intended to give a design a degree of robustness to improve the chances of survival when unforeseen events occur. Some parameters cannot be derated since they are functional elements of the component. For example, no voltage derating can be applied to Zener diodes, voltage clamps, and VDRs (voltage-dependent resistors) or varistors. The critical parameter that is controlled in these cases is temperature rise or power dissipation. Worst-case derating does not allow a direct estimate of failure rate or mean time between failures (MTBF). If the nominal and worst-case conditions are similar, some consideration may need to be given to further derating. Routine tests should not overstress any component. If the product specification requires a short circuit test on an output, the effect of this test on a power FETs Vds, for example, must not be to exceed the permitted voltage or avalanche energy level. If a short circuit test is required, it may have to be done at a line voltage that allows compliance with these derating requirements.

These requirements provide for some relief for low-energy leakage inductance spikes. A leakage inductance spike will occur in any circuit where current is switched and will always be resonant in nature.

When selecting a component, in general, if the component specifications do not specify a specific rating, then the part should not be allowed to operate in that mode: e.g. resistor pulse rating.

5.4.6 Ceramic Capacitor Cracks

Typically, cracks in ceramic chip capacitors can be caused by either thermal or mechanical stress. In the thermal situation, stress is due to a rapid change in temperature, which could be attributed to the reflow process, cleaning, wave soldering, or the rework process. The crack is due to the inability of the capacitor to relieve stresses during transient conditions.

Figure 5.3 Thermal stress crack. Source: Ansys-DfR Solutions.

Figure 5.4 Visible thermal stress crack. Source: Ansys-DfR Solutions.

The maximum tensile stress occurs at the end terminations, is electrically detectable, and is very difficult to find visually (Gulbrandsen et al. 2014).

Figure 5.3 shows what a typical thermal stress crack looks like.

Detecting a thermal crack visually is rare. Figure 5.4 shows a visually observable thermal stress crack. This occurred on an 1812-size capacitor, which is not recommended for environments where a high degree of thermal change will occur.

Variations in voltage or temperature will drive crack propagation while inducing a different failure mode: an increase in electrical resistance or a decrease in capacitance. Figure 5.5 shows the vertical crack under the termination, causing an open electrically due to the thermal crack.

Some things can be done to reduce this type of issue during the design portion of the DfM assessment. Avoid certain dimensions and materials:

- Maximum case size for SnPb: 1210.
- Maximum case size for SAC305: 0805.

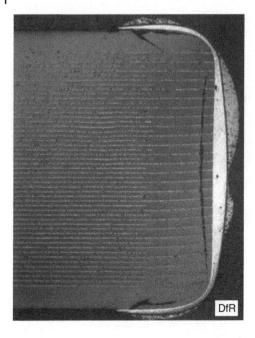

Figure 5.5 Vertical crack under termination. Source: Ansys-DfR Solutions.

- Maximum thickness: 1.2 mm.
- C0G, X7R preferred dielectrics.
- Adequate spacing from hand-soldering operations.
- Use the manufacturer's recommended bond pad dimensions or smaller, as smaller bond pads reduce the rate of thermal transfer.

Similarly, several things can be done in manufacturing to prevent this failure mode. The following are Ansys-DfR recommendations for a typical surface-mount technology (SMT) reflow process and wave soldering operations:

- Room temperature to preheat (maximum 2 to 3 degrees C/sec).
- Preheat to at least 150 C.
- Preheat to maximum temperature (maximum 4 to 5 degrees C/sec).
- Cooling (maximum 2 to 3 degrees C/sec).

It is recognized that the cool-down ramp conflicts with a profile from J-STD-020C (6 degrees C/sec). The faster rate can cause a quenching action that imparts stress to the solder joints. Make sure the assembly is less than 60 C before cleaning. For selective and wave soldering, maintain belt speeds toat a maximum of 1.2 to 1.5 m/min and eliminate cosmetic touch-ups.

Ceramic capacitors can also be cracked by excessive flexure of the circuit board under the capacitor, which can induce cracking at the terminations. The flexure can occur at depaneling, handling (i.e. placement into a test jig), insertion (i.e. mounting insertion-mount connectors or daughter cards), and attachment of board to other structures (plates, covers, heatsinks, etc.). Drivers for failure involve the distance of the capacitor from the flex point, its orientation on the circuit board, and the length of the component, as larger capacitors are more susceptible. From a DfM perspective, you can use the following guidelines:

- Avoid case sizes greater than 1206.
- Maintain 30–60 mil spacing from the flex point.
- Reorient parallel to the flex point.
- Replace with Flexicap (Syfer) or Soft Termination (AVX).
- Reduce the bond pad width to 80 to 100% of the capacitor width.
- Measure board-level strain (maintain below 750 microstrain; below 500 microstrain preferred for LF).

Low-profile surface-mount components can be impacted by mechanical shock, largely driven by board flexure. This situation can lead to pad cratering, intermetallic fracture, or even component cracking. Figure 5.6 shows examples of these failures. The upper image shows a crack at the ball grid array (BGA) package to the solder ball interface; the lower-left image shows a mechanically induced crack in a ceramic capacitor; and the bottom-right image is a pad-cratering crack under the BGA ball.

When evaluating a process for these failure modes, it is necessary to look beyond the common belief that the damage was caused by the product being dropped or mishandled. Excessive flexure can occur at multiple points in post-assembly operations. For example, flexural stress can occur at locations where heat sinks have been attached; where connector insertion forces may be excessive; where an in-circuit test (ICT) fixture has created too much localized force, causing board deflection; and during the depanelization operation, where poor support can lead to localized deflections of the PCB.

A DfM analysis can offer several mitigation techniques. For example, additional mount points to support the PCB may help. Also, using epoxy bonding, a thicker PCB, and underfill may enhance the mechanical

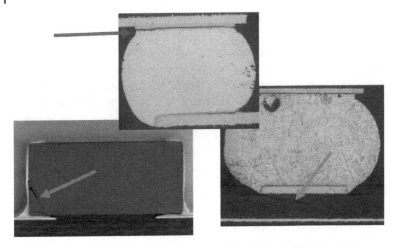

Figure 5.6 Mechanical shock failure modes. Source: Ansys-DfR Solutions.

strength of the assembly. Flexible terminations (on capacitors) may obviate any cracking of those devices; and corner staking, edge bonding, or underfill may benefit BGA attachment.

Pad cratering, which is cracking of the PCB laminate below a component, is typically due to a dynamic mechanical event caused by ICT, depanelization, connector insertion, or mechanical shock and vibration stresses. The drivers for this type of failure are finer-pitch components, more brittle laminates for LF applications, stiffer solder for LF, and the presence of a large heat sink. This failure mode is difficult to detect using standard X-ray, shear or pull testing, and die and pry methods.

Pad geometry can have an impact on pad cratering, as smaller pads result in higher stress under a given load. Also, soldermask-defined pads can provide additional strength. However, this approach moves the failure mode from pad cratering to intermetallic fracture.

Similarly, an ICT fixture (shown in Figure 5.7) can impose high localized stresses due to the pogo pins applying contact pressure in a small space and causing the board to flex. To reduce the pressures exerted on the PCB, the first and simplest solution is to reduce the probe pressure. If this can't be done, then using supports, fingers, or stoppers must be optimized to control the probe forces. However, this is difficult to accomplish, as the stoppers must be placed exactly under the pressure fingers to avoid the creation of shear points.

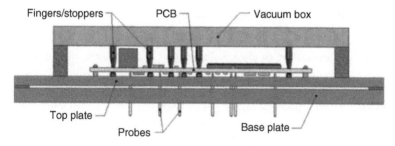

Figure 5.7 ICT fixture.

From a DfM viewpoint, several things can be done to mitigate this issue. For example, in the PCB layout, pad geometry may play a large part in conjunction with PCB thickness. It is necessary to keep the board flexure to less than 750 microstrain (500 microstrain would even be better). Other techniques include using a more compliant solder and different laminates.

5.4.7   Life Expectancies

Many parts are susceptible to long-term degradation in electronic designs. They require special consideration and include:

- Ceramic capacitors (oxygen vacancy migration)
- Memory devices (limited write cycles, read times)
- Electrolytic capacitors (electrolyte evaporation, dielectric dissolution)
- Resistors (if improperly derated)
- Silver-based platings (if exposed to corrosive environments)
- Relays and other electromechanical components
- Light-emitting diodes (LEDs) and laser diodes
- Connectors (if improperly specified and designed)
- Tin whiskers
- ICs (EM, TDDB, HCI, NBTI)
- Interconnects (creep, fatigue)
- Plated through-holes (PTHs)
- Solder joints

Each of these will be reviewed in more detail.

5.4.8 Aluminum Electrolytic Capacitors

Aluminum electrolytic capacitors are typically among the first components to wear out in an electrical assembly, making them the limiting factor in the lifetime warranty of many products. Once the voltage, capacitance, size, operating temperature, and ripple current are known, the capacitor lifetime can be optimized by choosing the most reliable manufacturer for the application. However, it can be difficult to identify the most reliable manufacturer without lifetime testing.

Aluminum electrolytic capacitor manufacturers state the endurance lifetime of their products in their datasheets, supported by testing that applies rated voltage and ripple current at the maximum rated temperature to the capacitor.

In service, electrolytic capacitors are rarely run at their rated values, so the endurance lifetime is not directly applicable. The industry uses a rule of thumb approach for lifetime: doubling life for every 10 C decrease below the rated temperature to estimate lifetime at lower temperatures. However, there is no explicit approach to estimate lifetime at lower applied ripple currents, which is also known to increase lifetime.

From a DfM perspective, it is useful to determine the life expectancy of the electrolytic capacitors as early in the design process as possible. The constant rate of electrolyte evaporation before thermal runaway is dependent on the external and internal temperatures of the capacitor. An increase in either temperature causes an increase in the evaporation rate. The external temperature is dependent on the ambient temperature. The internal temperature is impacted by the applied ripple current and frequency, where larger ripple currents and higher frequencies result in higher temperatures.

The best way to measure electrolyte evaporation is to monitor the volume of electrolyte in the capacitor over time. It is difficult to measure this directly. For electrolytic capacitor testing, it is assumed that at elevated temperatures, the only mass loss occurring from the capacitor is due to electrolyte evaporation. Therefore, tracking the change in mass of the capacitor throughout the test duration is a direct indicator of the change in electrolyte volume.

The impact of this primary wearout failure mechanism on the electrolytic capacitor lifetime is typically determined through a traditional test. The traditional test lifetime estimate is based on applying the rated voltage and ripple current at 120 Hz at the maximum rated temperature, which is typically 105 C. Most applications use capacitors at derated values; however, applying derated values significantly increases the lifetime, and test time, of the electrolytic capacitor. Running two traditional tests to compare capacitors with the same (or similar) rated voltage, capacitance, ripple current, can size, and maximum operating temperature can be time- and cost-intensive, especially when applying application conditions. A solution to this problem is to use an accelerated test approach, which takes a lesser, fixed amount of time regardless of the applied test conditions.

The accelerated life test for aluminum electrolytic capacitors consists of critical weight-loss testing and rate-of-weight-loss testing (Gulbrandsen et al. 2014). In critical weight-loss testing, the evaporation of electrolyte from the capacitor is accelerated by puncturing a hole in the can and baking it at rated temperature (typically 105 C). As the electrolyte evaporates, the internal resistance of the capacitor increases exponentially, just as in the traditional test. The equivalent series resistance (ESR) and weight loss are measured regularly during testing to determine the relationship between weight loss and ESR increase. Once ESR has increased to failure, typically defined in the datasheet as 200% from the initial value, the critical weight of electrolyte loss can be calculated. The critical weight is the amount of electrolyte loss needed to induce electrical failure via ESR.

In the second part of accelerated life testing, the rate of weight loss due to electrolyte evaporation is measured. This rate can be detected after 500 hours of testing and has been shown in traditional testing to stay constant. Using the results from the accelerated critical weight testing, the critical weight is calculated using the datasheet-defined increase in ESR as the failure criteria. The rate of weight loss is then extrapolated to this critical weight loss, and the time to failure for the test population of capacitors is determined. Confidence intervals on the data provide a range for the estimated time to failure. Results are specific to a given population of aluminum electrolytic capacitors and can vary by size, product family, manufacturer, facility, and date code.

5.4.9 Resistors

The DfM analysis performed should assess how well the resistors in the design are derated. The resistors should be aggressively derated regarding power dissipation, especially for the higher-power (1/4 watt and 1/2 watt) components. To maximize reliability, the following is recommended: derate to 60% of maximum applied voltage, 70% of steady-state power, 80% of peak power, and 20 C below peak operating temperature.

From a reliability perspective, if resistors with resistance equal to or greater than 1 megohm are used for sensing or calibration, there is a risk of intermittent resistance drift during operation. Use multiple lower-value resistors instead in these situations.

Metallization below the resistive element is usually made of silver. Silver is very susceptible to exposure to sulfur dioxide, which attacks the silver of the resistor and drives an open circuit. Figure 5.8 illustrates this issue.

The DfM analysis needs to assess the characteristics of the field environment and determine whether this failure mode is possible. There are other resistor metallizations available that don't have this issue, and they should be explored.

5.4.10   Tin Whiskers

Given the use of tin-plated components throughout the history of electronics, and with the transition to LF components, there is a need for a methodical approach to qualify and quantify the risk for a tin whisker to induce a failure. For each failure mode, existing knowledge of the behavior and physics of whiskering must be understood to demonstrate maximum whisker length, likelihood of whisker breakage, value of a conformal coating, and reliability prediction based on a Monte Carlo simulation. Performing a whisker risk assessment as part of a DfM review is highly recommended (Caswell 2015c).

A *risk assessment* is a practice of determining if the actual reliability will meet or exceed the desired reliability and the consequences if these performance goals are not met. The actual reliability of a product is driven by the failure modes that can initiate a failure and the materials, design, and environment that interact to drive failure processes.

Performing a risk assessment requires a review of the failure modes that can initiate due to tin whiskering, the materials and design existent within the product, and the expected operating environments.

Failure Modes   Three primary failure modes can be initiated due to the presence of tin whiskers. Static contact occurs when a tin whisker grows

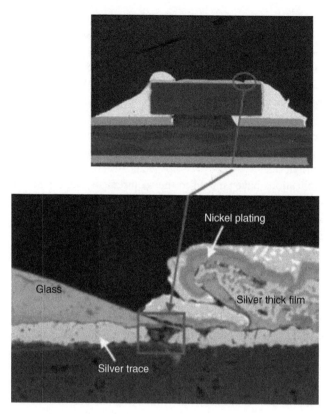

Figure 5.8 Resistor damaged by sulfur dioxide. Source: Ansys-DfR Solutions.

and makes a physical connection with an adjacent conductor. Debris is when a whisker becomes dislodged and has the potential to provide a conductive bridge to other locations within the product. Both failure modes create a leakage path, inducing an electrical short. The third potential failure mode is radiated emissions in high-frequency applications, typically exceeding 5 GHz.

A critical aspect of this risk assessment is the differentiation between matte tin and bright tin plating. This terminology was initially used by plating chemists, who focused on differences in plating chemistry and visual appearance. Bright tin is usually plated in sulfate baths, while matte tin is plated in stannate baths.

More recently, attempts were made to provide a more quantitative and physical justification for the terms *matte* and *bright* tin. Since this effort was still primarily based on visual appearances, the results provided more of a general range than an exact value. These ranges consisted of organic content and grain size. The brightening agents in sulfate baths generate more co-deposited organic content in bright tin. As an example, one plating company quantifies the organic content for its bright tin plating to be approximately 0.15%. The organic content also influences grain size. Matte tin has a grain size approximately an order of magnitude greater than that of bright tin.

Several publications have pointed to the requirement that a whisker must grow in the correct orientation to contact an adjacent conductor (Galyon et al. 2005; Lee and Lee 1998; Galyon 2005). The required angle of growth for the solder terminal is highly variable due to the round surface of the outer ring. As a result, while acknowledging that even whiskers that grow the length of the separation distance may not come into physical contact with adjacent conductors, a quantification of this risk is difficult.

Using conformal coating as a barrier to tin whiskering has been assessed by several organizations (Kadesch et al. 2000; Kadesch and Brusse 2001; Han et al. 2012; Woodrow and Ledbury 2006), with the most quantitative information provided by NASA. There are two primary application methods when using conformal coating. The application of conformal coating over the tin whisker-producing surface can reduce exposure to humidity, a potential whisker accelerant, and has been shown to constrain whisker growth if the coating is thick enough. However, since there is some uncertainty about the application of conformal coating over the solder terminal, the more applicable approach is conformal coating of the surface where the whisker will make physical contact.

As such, the DfM assessment needs to look at the pitch of all the parts in the BOM and their surface finishes, and determine whether conformal coating may be used.

5.4.11 Integrated Circuits

During a DfM analysis, it is necessary to identify the gate geometries for all the ICs in the product. Doing so provides insight into whether devices can meet the long-term reliability requirement for the product.

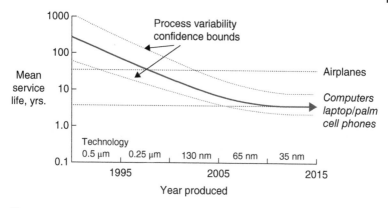

Figure 5.9 IC wearout concern.

Smaller geometries in parts have created a new failure mode: IC wearout. Figure 5.9 shows how using smaller geometries has significantly reduced the mean service life for several products.

5.5 Printed Circuit Board Topics

To ensure that a design is manufacturable, many checks are performed at the PCB design stage.

5.5.1 Laminate Selection

Discussion of Tg

The documentation for typical boards should specify a high glass transition temperature (Tg) laminate when dealing with the temperatures involved with LF assembly. A Tg greater than 180 C is recommended. IPC 4101 is the guideline for raw PCB materials. It contains six slash sheets for LF-capable materials, which fall into three main classifications. However, Tg alone is not a good indicator of a materials' suitability for LF assembly. While a material with higher Tg is more thermally resistant, new materials with high or low Tg can be LF-capable. One of the major differences between LF FR4 and standard FR4 is the Td (decomposition temperature). LF FR4 materials have a Td around 350 C. It is also important for a material to be able to achieve a T288 (time to delamination) value of at least 5 minutes. Boards should be subjected to multiple LF reflow cycles to verify resistance to delamination and blistering. So, specify Tg, Td, and T288 when selecting PCB laminates.

5.5.2 Surface Finish

The selection of the PCB surface finish could be the most important material decision made for the electronic assembly. The surface finish influences the process yield, amount of rework needed, field failure rate, ability to test, scrap rate, and cost. You may be led astray by selecting the lowest-cost surface finish, only to find that the total lifetime cost is much higher.

The surface finish should be selected with a holistic approach that considers all essential aspects of the assembly (Caswell 2013a):

- Cost sensitivity
- Volume of product (finish availability)
- SnPb or LF process
- Shock/Drop a concern?
- Cosmetics a concern?
- User environment (corrosion a concern)?
- Fine pitch assembly (less than 0.5 mm)?
- Wave solder required (PCB thickness greater than 0.062 in.)?
- ICT required?

Figure 5.10 shows the various surface finishes and their respective characteristics.

5.5.3 Discussion of Different Surface Finishes

Hot air solder leveled (HASL)

LF HASL is increasingly the solderability plating of choice, as it has excellent wettability/solderability and shelf life. It is also robust in multiple reflow cycles and produces strong solder joints.

The issue with this finish is that it has poor to fair flatness/coplanarity compared to other LF finishes (but improved coplanarity over SnPb HASL). The thickness varies across the different size footprints in a design. The air-knife pressure, pot temperatures, and nickel content all have an impact on planarity. And as the HASL process exposes/dips the

Surface finish	Cost	Corrosion res.	ICT	Hole fill	Fine pitch	Cosmetics	Comments
Immersion silver	Low	Poor	Good	Mod	Good	Poor	Good surface finish for soldering and testing. Tarnish and creep corrosion are the weaknesses.
Organic surface protectant	Low	Mod	Poor	Poor	Good	Mod	Requires pasting of test pads/vias. Difficult to achieve LF hole fill, especially on >0.062 boards.
Lead-free hot air solder level	Mod	Good	Good	Good	Mod	Good	Phenolic laminate recommended. New equipment required. Flatness is better than SnPb HASL.
Immersion tin	Mod	Good	Good	Mod	Good	Mod	Solderability/hole-fill may be a problem on double-sided PCBs. Shelf life.
Electroless nickel immersion gold	High	Mod	Good	Good	Good	Good	Galvanic-driven creep corrosion can occur if Cu is exposed. Ni-Sn interface is brittle with LF. Black pad issues remain.

Figure 5.10   Surface finishes.

board to hot liquid LF solder, there is also some risk of thermal damage to the board.

For optimum reliability, a thickness of 100 microinches is recommended. Lower thicknesses can result in exposed intermetallic.

Electroless-nickel-immersion-gold (ENIG)

Since gold readily dissolves in solder and does not tarnish or oxidize, it is an excellent choice for a surface finish. But gold cannot be directly plated onto copper since copper diffuses into gold. Diffusion allows the copper to reach the surface and oxidize, which reduces solderability. As such, a nickel layer is used to serve as a barrier between the copper and gold. The gold protects the nickel from oxidizing.

There are still potential issues with this surface finish, particularly with many overseas fabricators. One issue is the resurgence of black pad, which is impacted by phosphorus content. This failure mode impacts solderability and mechanical strength and can affect a random number of pads. The appearance of black pad is shown in Figure 5.11.

In general, ENIG is less robust than organic solderability preservative (OSP) for mechanical shock, so environments where the product will be used must be known and understood.

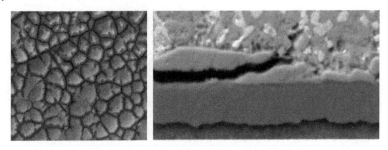

Figure 5.11 Black pad images. Source: Ansys-DfR Solutions.

Electroless-nickel-electroless-palladium-immersion-gold (ENEPIG)

ENEPIG is a surface finish where electroless palladium is incorporated between the nickel and the gold layers to act as a barrier that further reduces copper diffusion and grain boundary corrosion of the nickel surface. The palladium is absorbed into solder without high-phosphorous regions, which becomes an oxide-free nickel layer for optimum solderability. This finish is also viable for wire-bonding applications. Although it is more expensive than ENIG, ENEPIG is less costly than electrolytic nickel, electroless palladium, and immersion gold.

Immersion Silver (ImAg)

ImAg finish is a single-material system that is specified by IPC-4553. The thickness of the finish is defined as a minimum thin thickness of 2 microinches to a thicker 5 microinches. The material has good flatness and coplanarity, has a good shelf life if properly stored and protected, has reasonable oxidation resistance (again if stored and handled properly), has good wettability and solder reflow performance, and works well in test. This finish is also often chosen due to its low cost.

However, this finish has several negative attributes as well. An ImAg finish is susceptible to galvanic trace etching if the PCB is rinsed but not completely dried. It is also susceptible to planar microvoids, which translate into an issue known as *champagne voiding*. The root cause of this issue is still under debate. The voiding could be caused by plating that is too thick, an insufficient peak temperature in processing, trapped gas, oxidation of the surface finish, and/or some form of intermetallic reaction formation. Tarnish on the surface leads to solderability problems, and significant degradation has been observed when the finish is exposed

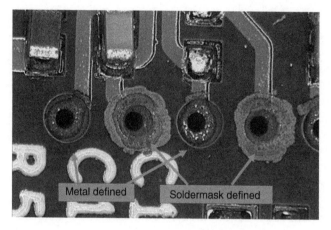

Figure 5.12   Silver creep. Source: Ansys-DfR Solutions.

to corrosive gases. It has been shown that a one-day test with mixed flowing gas (MFG) Class II conditions equates to 6 months in the field at light industrial conditions. So, this test can be used to evaluate a product's susceptibility in each corrosive gas environment.

Another ImAg finish issue is creeping corrosion, where the silver migrates and can cause shorting between traces that are susceptible to leakage current. This issue seems to be more of a problem in certain industries like paper mills, wastewater treatment, petrochemical plants, and rubber manufacturing. This issue is shown in Figure 5.12.

Organic Solderability Preservative (OSP)

OSP is an organic coating that is processed onto copper pads to protect the pads before soldering. It is a good surface finish for operation in LF applications, and it provides a flat surface for assembly. However, it has issues with coating into PTHs and has a short shelf life when stored before assembly. Some important parameters are hole diameter, hole aspect ratio, and wetting force, as solder will fill only if it is molten. If prior heat and flux have removed the OSP from these holes, poor wetting may result due to copper oxidation buildup. This could lead to single-sided solder joints (poor hole fill) on through-hole parts, which is not as reliable as a completely filled hole. OSP also presents a problem with ICT, as it is difficult for the probes to penetrate the coating or subsequent oxidation.

Immersion Tin (ImSn)

ImSn is a single-material system specified by IPC-4554. The standard thickness is 40 microinches, but some companies specify up to 65 microinches. It is a low-cost material with excellent flatness. Like the other finishes, it also has challenges. There are environmental and health concerns when processing ImSn due to the use of thiourea, a known carcinogen. Also, the finish itself is at risk of tin whiskering.

If the ImSn finish is insufficient, solderability decreases during storage and after the second reflow due to the growth of an intermetallic compound (IMC) through the thickness. If an oxide that is greater than 5 nm builds up on the surface, a solderability problem usually occurs. Exposure to humidity greater than 75% accelerates oxidation through the creation of tin hydroxides. As such, PCBs need to be packed for storage either in nitrogen or in sealed bags with desiccant and humidity indicator cards (HICs). Cleanliness of the board is paramount as contaminants can break down the self-limiting nature of tin oxides and impact good solderability.

5.5.4 Stackup

As the speeds and complexity of electronic components have continued to climb, the importance of PCB stackup analysis in the overall design process has become even more important. The common practice of leaving the vital PCB stackup design in the hands of the PCB fabrication company's fabrication engineers carries significant hazards. Often, the only criteria given to a fabricator are the overall number of layers, desired total thickness, and expected impedance of the transmission lines (if any). A much better approach is for the design team to provide the fabricator with all the information necessary for creating a PCB that is fully compliant with the specifications.

The term *stackup* refers to positioning and types of layers in a multilayer board. The design of PCB stackup involves determining the number of signal layers needed to route the board and the ground/power planes for adequate power distribution. The layers of the board should be arranged so that each signal layer is referenced to a continuous plane. Furthermore, stackup should include parallel (adjacent) power and ground layers for enhancing capacitive decoupling and reducing EMI. PCB stackup also strongly influences the impedance of a power distribution system (PDS).

Among the demands placed on stackup design and analysis are:

- Providing enough signal layers to allow successful routing of all signals to achieve signal integrity requirements.
- Specifying copper thickness for each layer that meets the conductivity requirements of signals and power and, at the same time, is manufacturable.
- Providing enough power and ground layers to meet the different power/ground requirements.
- Specifying trace widths and dielectric thicknesses that allow impedance targets to be met.
- Ensuring that the spacing between signal layers and their adjacent planes is thin enough to minimize crosstalk.
- Specifying laminate/dielectric materials that are economical to manufacture and readily available. Material availability from vendor to vendor varies. The materials should be evaluated for the life of the design to prevent any future stackup changes.
- Avoiding the use of expensive techniques such as blind and buried vias and build-up processing when possible.
- Providing for prototype manufacture in one factory or country and production manufacture in another factory or country.
- Restriction of Hazardous Substances (RoHS) and halogen-free compliance.

When designing multi-layer boards, there are five objectives to achieve:

- A signal layer should always be adjacent to a power or ground plane.
- Signal layers should be tightly coupled (close) to their adjacent planes.
- Power and ground planes should be closely coupled.
- High-speed signals should be routed on buried layers located between planes. This way, the planes can act as shields and contain radiation from the high-speed traces.
- Multiple ground planes are very advantageous since they lower the ground (reference plane) impedance of the board and reduce common-mode radiation.

Eight layers is the fewest that can be used to achieve all five of these objectives. Another desirable objective, from a mechanical point of view, is to have the cross-section of the board symmetrical (balanced) to prevent warping as a result of the fabrication process. For example, on an eight-layer board, if layer 2 is a plane, then layer 7 should also be a plane, resulting in symmetrical, or balanced, construction.

5.5.5 Plated Through-Holes

PTHs, also known as plated through-vias (PTVs), are holes drilled through multilayer PCBs that are electrochemically plated with conductive metal (typically copper). These plated holes provide electrical connections between layers. Because PTHs are metallurgically bonded to annular rings on the top and bottom of the PCB, they act like rivets and constrain the PCB. This constraint subjects the PTH to stresses when the PCB experiences changes in temperature. Over time, the PTH experiences fatigue and eventually fails due to crack propagation. PTH fatigue is the circumferential cracking of the copper plating that forms the PTH wall. It is driven by differential expansion between the copper plating ($\approx$17 ppm) and the out-of-plane CTE of the printed board ($\approx$70 ppm). Figure 5.13 shows an example of this failure mode. PTH fatigue is influenced by a number of drivers, including temperature range, PTH diameter, PTH copper plating thickness, copper plating material properties, PCB thickness, PCB out-of-plane material properties, and quality of the copper plating. Calculate time to failure using the industry-accepted model published in IPC-TR-579, "Round Robin Reliability Evaluation of Small Diameter Plated Through Holes in Printed Wiring Boards." Life calculation for PTHs subjected to thermal cycling is a three-step process involving a stress calculation, strain-range calculation, and iterative lifetime determination.

5.5.6 Conductive Anodic Filament (CAF) Formation

CAF is the migration of copper filaments within a PCB under an applied bias. When the copper filaments bridge adjacent conductors, they can cause an abrupt, unpredictable loss of insulation resistance. The migration almost always occurs along the weave of the glass fiber bundle embedded within the laminate and prepreg layers that make up a rigid PCB. Due to the presence of drilling damage and the large surface area of copper, the dominant failure site for CAF is between adjacent PTHs. CAF is influenced by electric field strength, temperature, humidity,

Figure 5.13 PTH failure.
Source: Ansys-DfR
Solutions.

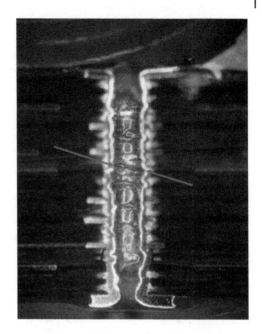

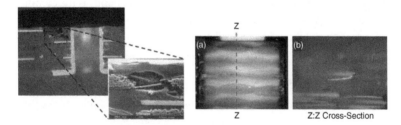

Figure 5.14 CAF examples. Source: Ansys-DfR Solutions.

laminate material, soldering temperatures, and the presence of manufacturing defects. Software is available that benchmarks the printed board design and quality processes to industry best practices, including the wall-to-wall distance between the PTHs along the orthogonal axes, degree of overlap, and frequency and type of qualification performed to assess CAF performance. Figure 5.14 provides examples of how a conductive anodic filament can result in a shorted PCB.

Table 5.2 Copper weight.

Copper weight (ounces)	Copper thickness (mils)
1/2	0.7
1	1.4
2	2.8

5.5.7 Copper Weight

The weight of the copper on each layer of a PCB is typically specified in ounces. The most common weight is 1 ounce, which is determined by calculating 1 ounce of copper spread evenly over 1 square foot of area. The thickness then becomes 1.37 mils or 0.0347 millimeters. In typical PCB fabrication, one-ounce copper is used for the inner layers, and half-ounce copper (then plated up) is used for the outer layers. If the design has a higher current flow, the designer has two options: either make the traces wider or make them thicker. Due to space constraints with trace widths, when thicker copper is used, the cost of the PCB typically increases. The minimum trace and the copper thickness go hand in hand, as the etching process has an impact on the cross-section. Thicker copper requires wider traces. Table 5.2 illustrates typical copper weights and thicknesses.

5.5.8 Pad Geometries

IPC-7351 and its subsets provide the best resource regarding footprints (land patterns) for different device types. The selection of footprints during the design stage has a profound impact on the reliability of the product. As such, they need to be part of the DfM analysis. They are an integral part of the circuit board and are subject to the manufacturing capabilities and tolerances of the fabrication facility and, ultimately, the assembly facility.

Land patterns or footprints need to be assessed to ensure viable and reliable solder joint formation, soldermask clearance from the reflowed solder joints, and clearance between components and the solder joint bondline thickness. Some designers feel that larger pads provide more reliable connections, but this is not always true. As the solder runs from

Figure 5.15
QFN bondline.

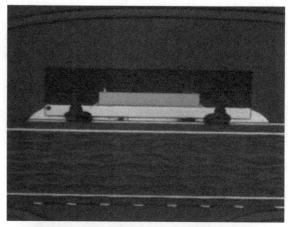

the bondline to the edges of the pads, the result is thinner, less reliable, bondlines.

Concerning footprints, the area where the most variability has been seen is with quad flat-pack no-lead (QFN) packages. Stencil thickness and aperture design are crucial for the manufacturability of these devices. The reliability objective for the bondline of these devices (and any bottom-terminated device) is to have a solder joint bondline (thickness) between 2.5 and 3 mils thick. To achieve this requires IPC footprints combined with a 5 mil thick stencil. Figure 5.15 shows the correct appearance of a QFN solder joint and the requisite thickness of the bondline.

Another issue with QFNs is the utilization of thermal vias in the thermal pad. If these vias are not filled before soldering, the solder paste applied to the thermal pad runs down the vias and starves the solder joint. The surface tension of the solder on the thermal pad pulls the package down closer to the surface of the PCB and pushes the solder out on

Figure 5.16 QFN I/O pad and thin bondline.

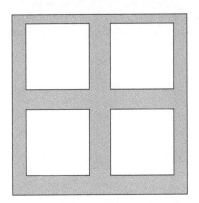

Figure 5.17 Windowpane stencil structure.

the I/O pins. The result is a bulbous solder joint with a thin bondline, as shown in Figure 5.16 (Caswell and Tulkoff 2014).

Proper solder paste printing on a thermal pad is also an issue. A windowpane print is recommended to enable 65 to 80% coverage of the thermal pad. The minimum spacing between print zones is 6 mils to permit the escape of gases associated with flux and solder paste additives. Figure 5.17 illustrates the windowpane stencil structure.

5.5.9 Trace and Space Separation

The best method to achieve good manufacturability for the board layout with regard to trace widths and spaces between traces is to follow the guidelines of IPC-2221, the accepted industry design standard. IPC-9592 can also be used. The two standards use slightly different methods for calculating the spacing between traces as a function of the applied voltage.

If products have a safety requirement, then the creepage and clearance requirements of a UL-IEC specification are mandatory.

5.5.10 Non-Functional Pads

There is an ongoing debate regarding the influence of non-functional pads (NFPs) on PCB reliability, especially as related to barrel fatigue on PTVs with high aspect ratios (Binfield 2015a). The industry has indicated that most suppliers do remove unused/non-functional pads. No adverse reliability information was noted with the removal of unused pads; conversely, leaving them can lead to an issue called *telegraphing*. In either case, the primary reason given by suppliers for removing them was to improve the respective fabricators' processes and yields. Companies that remove unused pads do so primarily to extend drill-bit life and produce better vias in the boards, which they consider the primary reliability issue. For those that keep the unused pads, the primary reason given was that they believe it helps manage the Z-axis expansion of the board due to the CTE stresses. However, with the newer LF assembly materials, the Z-axis CTE concern seems to have abated. From a DfM perspective, it is necessary to consider this design issue.

5.5.11 Shipping and Handling

From a DfM perspective, controlling how PCBs are shipped from the fabricator to the user is essential. PCBs should be shipped in sealed packaging with desiccant and humidity indicator cards, and they should be stored in a temperature- and humidity-controlled environment.

5.5.12 Cleanliness and Contamination

Cleanliness and contamination issues are believed to be primary drivers of field issues in electronics today. Contamination induces corrosion and metal migration (electrochemical migration [ECM]).

Intermittent behavior of a product tends to produce no-fault-found (NFF) returns, and intermittent failures are extremely difficult to diagnose. These failures can be driven by the self-healing behavior of contamination-induced defects. Dendritic growth can create shorts between conductors that burn off as heat is generated. The unit will then pass, but it can fail again if the dendrite regrows. This pattern of pass and fail can recur if there is fuel for dendritic growth (Tulkoff 2013; Binfield 2015b).

Cleaning and contamination failure modes have been observed on batteries, liquid crystal displays (LCDs), PCBAs, wiring, switches, etc. The failures have increased as components continue to get smaller and as the spacing between the bottom of these parts and the surface of the PWB gets smaller. Thinner bondlines make it almost impossible to clean under low-profile parts unless some form of saponifier is used. Continued reductions in pitch between conductors make future packaging more susceptible, as does increased use of leadless packages (QFN, land-grid array, etc.). Using leadless packages results in a reduced standoff or bondline thickness. The efficiency of cleaning under low-profile components is also reduced, which leads to increased concentration of contaminants underneath them.

In today's global economy, increased product sales to countries with polluted and tropical environments have driven increased ECM due to ambient humidity conditions. LF and smaller bond pads also require more aggressive flux formulations, exacerbating cleanliness issues.

With greater use of no-clean flux formulations, it is now becoming necessary to requalify certain manufacturing processes. Design and material combinations can result in elevated weak organic acid concentrations, which leads to dendritic growth. Changes to the water temperature, saponifier, and impingement jets in cleaning systems may be necessary. The smaller pads on smaller-geometry parts also require less solder paste. Less solder paste means less flux. This drives the use of higher-activity fluxes, which in turn impacts cleanliness. Figure 5.18 illustrates this concept.

With the changes in component technology and continued miniaturization of components, cleanliness has evolved into more of an issue with ECM. In the classical formulation, ECM is a four-step surface process: the creation of a pathway between two conductors, electrodissolution of metal, ionic migration, and electrode position. The creation of a path is either moisture adsorbing or condensing on the surface. Electrodissolution requires a metal anode that ionizes (oxidizes) under bias. Once oxidized, metal ions migrate to and deposit on the cathode. The deposited metal can grow to completely span the conductors or at least decrease the effective conductor spacing.

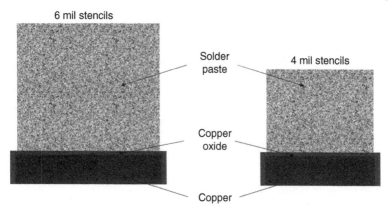

Figure 5.18 Solder paste volume change. Source: Ansys-DfR Solutions.

All that is needed for leakage current is a conductive pathway, regardless of whether metals are oxidizing and migrating and their rates of oxidation and migration. Some leakage current is passed through any liquid medium with dissolved ionic contamination spanning conductors under bias.

Conductive anodic filament (CAF) formation typically refers to the same process as ECM but occurs within the laminate between internal copper layers or PTHs. The pathway that CAF follows is often formed by delamination or hollow glass fibers (note that refinement of the glass fiber extruding process has largely eliminated hollow fibers).

In the electronics industry, tin, lead, copper, and silver are the most common metals involved in ECM and CAF. Some metals such as gold and steel are significantly more resistant to electrochemical oxidation. So, where are the contaminants coming from in the board fabrication process, and how can they be mitigated during DfM assessment? Table 5.3 illustrates the process operations and what contaminants might come from that operation.

At a minimum, PCB manufacturers should clean PCBs:

- Immediately before the application of solder resist.
- Immediately after the application of any solderability plating, after HASL.
- After surface finish processing regardless of whether it is ENIG, ImSn, or ImAg.

Table 5.3 Sources of contaminants.

Ion	Possible sources
Cl	Board fab, solder flux, rinse water, handling
Br	Printed board (flame retardants), HASL flux
Fl	Teflon, Kapton
PO^4	Cleaners, red phosphorus
SO^4	Rinse water, air pollution, paper and plastics
NO^4	Rinse water
Weak organic acids	Solder flux

Some PCB manufacturers also perform a final clean before shipment. Caution: fabricators should not substitute this cleaning for cleaning after solderability plating, as residues from plating operations can become more difficult to remove with time delay.

At the assembly level, cleaning requirements are driven by J-STD-001, which mandates 10 μg/in.^{2} (1.56 μg/cm^2) for ROL0 or ROL1 fluxes (others are based on limits established by the user).

Section 6.1 of J-STD-001

Assemblies should be cleaned after each soldering operation so that subsequent placement and soldering operations are not impaired by contamination.

Some guidance is provided by two handbooks:

1. "Guidelines for Cleaning of Printed Boards and Assemblies," IPC-AC-62A (2012)
2. "Aqueous Post Solder Cleaning Handbook," IPC-CH-65B (2012)

From a DfM perspective, it is necessary to understand the solder paste flux chemistry, cleaning process parameters (e.g. batch vs. in-line), cleaning solution chemistry (including saponifier), and cleaning temperature, and where in the process cleaning is implemented. After cleaning, the analysis should provide insight into whether the cleaning process is providing good results, or it should identify what needs to be changed to make the process effective.

Some actions that can be taken in the design to enhance the cleaning operation are as follows:

- No traces under components (most common)
- Smaller bond pads (lifts the component)
- Thinner soldermask (increases standoff)
- Components in parallel (prevents blocking of cleaning solution)
- Board cutouts (primarily for components with larger chips)

5.6 Process Materials

Many process materials are involved in electronics manufacturing. The following details some of the materials that should be assessed during a DfM analysis.

5.6.1 Solder

Solder is one of the most prominent process materials. There are significant differences in defects per million (DPMM) for solder joints depending on whether the joints were created by reflow, wave, or hand-soldering operations. In general, reflow soldering produces fewer defects than wave or selective soldering, which in turn produce fewer defects than hand soldering. Table 5.4 illustrates this concept.

5.6.2 Solder Paste

The DfM review must also assess the reflow profile to ensure that it is optimized for the assembly being soldered, solder paste chosen, and overall set of parts and materials being implemented.

Start with the paste manufacturer's recommendations. Select a ramp and soak vs. straight ramp preheating profile. Ramp and soak (soak period just below liquidus) is more common and more forgiving: it allows flux solvents to fully evaporate and activate to deoxidize the surfaces to be soldered, and the temperature is able to equalize across the entire assembly. However, if it lasts too long, the flux may be consumed, resulting in excessive oxidation. And the flux may become volatile, producing solder balls or voiding defects. A straight-line preheating

Table 5.4 Soldering process DPMM.

Soldering process	DPMM	
	Standard	Best in class
Hand	5000	N/A
Wave	500	20–100
Reflow	50	<10

profile is faster and causes less thermal damage to materials but is more susceptible to defect and quality variation. It also does not work as well on complex, dense assemblies.

Establish the correct peak temperature and time above liquidus (TAL), which is a balance between being hot enough for long enough to achieve good, consistent solder wetting and bonding for proper joint formation across the entire assembly. After this, proceed as quickly as possible to prevent thermal damage to the components and board and to prevent excessive copper dissolution and excessive intermetallic growth.

The cooling rate is equally important as it affects the microstructure and bulk intermetallics. A faster cooling rate produces a finer, stronger microstructure and limits intermetallics; but overall throughput is determined by the board size/complexity and the oven's heat transfer capabilities. A good rule of thumb is 2 to 3 degrees C/second ramp up and down rate.

5.6.3 Flux

Fluxes are composed of several components depending on the type. Because liquid fluxes (wave solder and hand-solder fluxes) have a relatively simple requirements list (apply, react with and dissolve metal oxides, and leave residues that are non-reactive or can be easily cleaned), the chemical components tend to be straightforward. The constituent with the largest percentage by volume is a solvent or solvent blend. The solvent is the first substance to evaporate during preheating before the molten solder bath. It dissolves and stabilizes the other ingredients. Water and organic alcohols are common solvents for water-soluble, liquid fluxes (Caswell 2013b).

The constituent with the second-highest volume percentage is called the *vehicle*. The vehicle is usually a co-solvent with a higher boiling point than the solvent. It keeps the ingredients and reaction products dissolved during preheating and soldering after the solvent has evaporated. Vehicles include glycols and polyglycols.

The remaining components are the active ingredients followed by proprietary additives. The active ingredients remove surface oxides. For water-soluble fluxes, they are either weak organic acids like succinic, glutaric, and adipic acids, or halide-based, common sources of chloride or bromide. The proprietary additives control other factors such as wettability, testability, and encapsulation of residues.

For pastes, the constituents are similar, except the vehicle is not present and the number of additives is higher. Generally, pastes are more complex, and ingredients are closely guarded trade secrets. The metal/flux mixture must have precise physical and chemical properties. The flux must be tacky enough to hold components in place, and the flux-to-metal ratio must be optimized to maintain the viscosity and shape of the paste during and after printing. The paste should spread easily at reflow temperatures, wetting the entire pad and component lead, termination, or pad.

Unlike liquid fluxes, pastes are exposed to high temperatures for significantly longer times. Because of this extended exposure time, pastes usually have several temperature ranges at which some ingredients evaporate and others fully activate. This allows the various ingredients to continue to remove oxides and prevent new oxide growth on the metal surfaces throughout the entire profile until the molten solder has completely wetted the pad and component. The metal melting point along with the boiling points and regions of high activity in the flux constituents determine the recommended reflow profile.

Because halides are strong electrolytes and inorganic, they remain active through the entire profile and do not evaporate. Weak organic acids are weak electrolytes and ideally become benign through evaporation and several decomposition pathways.

Flux before wave soldering is applied by either spray or foam, while solder paste is usually screen printed. These processes can tightly control the amount of flux present during soldering. Hand soldering generally has a wider distribution of flux volume.

Fluxes are designed to break down at soldering temperatures, ideally leaving behind minimal and benign residues. Solvents evaporate almost entirely; some may become trapped in the molten metal, forming voids in the metal after solidification. Weak organic anions mostly break down chemically or evaporate. Oxidation, esterification with solvent, complexing with metal ions, and possibly polymerization are all potential results. Most of these products should be benign because they are either insoluble in aqueous solutions or lose the structural ability to dissociate and hence lose their acidic property. Small amounts of unreacted acids can be surrounded and trapped by benign, insoluble molecules, making them less likely to be dissolved in water and reforming the acid.

During a DfM review, it is vital to understand the attributes of the fluxes in the appropriate manufacturing operations and evaluate whether they will present a problem once the product has been in the field for a while.

5.6.4 Stencils

The screen-printing operation in a typical SMT manufacturing process is the operation most responsible for subsequent defects. As such, correctly printing the correct volume of solder paste onto the corresponding surface-mount pads on the circuit board is critical. Achieving the correct volume is primarily a function of the thickness and apertures of the stencil Caswell 2014).

The focus of the screen-printing operation is to apply the correct volume of solder paste consistently on the correct pad locations on the PCB. Numerous manufacturing and design variables can impact a manufacturer's ability to do this. These variables typically translate into tolerance stackup issues, where the stencil and pads don't properly align even though they are made from the same design data.

Stencils are manufactured in several ways. They may be laser cut, chemically etched, electroformed, or, more recently, nano-coated.

To design a good SMT stencil, a systematic approach should be followed. This approach consists of four steps: a CAD or Gerber review, a BOM review, a specification review, and a PCB fabrication review:

- CAD or Gerber review: CAD or Gerber data is required because this information is used to make the stencil and define the apertures for each component.

- BOM review: Analyzing the BOM plays a very important role in the stencil design process. Information in the BOM helps to define the type of stencil technology required and the thickness of the stencil, and also provides insight into the reductions in apertures for certain packaging types.
- Specification review: Some packages have special requirements for stencil design (e.g. QFNs thermal pads). Some companies also have stencil design guidelines, and a review approach will help the designer incorporate these elements.
- PCB fabrication review: Optional, but the most important step. A solder sample is not always available to the stencil designer, but it provides information that is not included in CAD, Gerber, BOM, or standard guidelines. The information from a sample PCB permits an assessment of the variation in dimensions of the PCB pads from the Gerber data, which may require a stencil being fabricated from the surface image of the PCB.

The volume of solder paste applied to a pad can be determined by the area ratio, which is defined as the ratio between the aperture walls and the PCB pad with a given aperture design. The ratio determines the percentage of solder released on the PCB pad. IPC standards suggest that good solder release occurs at a ratio greater than or equal to 0.66. The area ratio is length times width over 2T times length plus width, where T is the thickness of the stencil.

Area Ratio Formula:

$$\frac{Length \times Width}{2T \times (Length + Width)} \tag{5.1}$$

The DfM review should assess these parameters and make sure they are followed for optimum manufacturability.

5.6.5 Conformal Coating

Conformal coating is applied to circuit cards to provide a dielectric layer on an electronic board. This layer functions as a membrane between the product and the end-use environment. With this coating in place, a circuit card can withstand more moisture by increasing the surface resistance or surface insulation resistance (SIR). With a higher-SIR board, the risk of problems such as cross talk, electrical leakage, intermittent signal losses, and shorting is reduced (Masterbond 2018).

This reduction in moisture exposure also helps reduce metallic growths called dendrites along with corrosion and oxidation. Conformal coating also serves to shield a circuit card from dust, dirt, and pollutants that can carry moisture and may be acidic or alkaline.

In today's manufacturing environment, with no-clean fluxes used on many products, there are concerns about applying conformal coating over flux residues. Many suppliers don't recommend doing this, stating that residues can reduce adhesion, potentially resulting in delamination, and that the coating-over-residue environment creates micro-condensation conditions that can be more detrimental than using no conformal coating at all.

The application where the coating will be used must also be considered. For example, the automotive industry specifies conformal coatings to protect circuitry from gasoline vapor, salt spray, and brake fluid. The use of electronic systems in vehicles is increasing rapidly, and as such, using conformal coatings is becoming vital to ensure long-term reliability. Conformal coatings are used in applications both under the hood (e.g. engine management systems) and in passenger compartments (e.g. onboard computers).

Aerospace industry products, with their high-reliability requirements, are also viable applications for conformal coatings. The environmental requirements of the aerospace industry, where rapid compression and decompression can affect the performance of circuitry, necessitates the use of conformal coatings. The coatings are used in both pressurized and depressurized areas.

Both freshwater and saltwater environments attack electronic circuitry. Conformal coatings are ideal for the protection of equipment used in both environments. These applications range from under the dash in high-performance boats to exterior equipment used on larger maritime systems.

Similarly, in the medical industry, there are numerous areas where a conformal coating is required for environmental protection: tool protection while in storage to prevent corrosion; pacemakers; and anywhere it is vital to ensure continuous performance. Coatings also prevent an attack from the human body itself.

Several types of conformal coating materials are available. Selecting a coating requires an evaluation of variables since each coating works best in specific environments.

Silicone

Silicone conformal coatings are most widely used in high-temperature environments due to their innate ability to withstand prolonged exposure to higher temperatures (200 C). This attribute has made them the primary choice for underhood automotive applications. They can also be applied in thicker films, making them useful as a vibration-dampening/isolation tool if the coated assembly is to be placed in a high-vibration environment. Rework of silicone-coated assemblies can sometimes be difficult due to their chemical resistance. Unlike acrylic and polyurethane coatings, silicone does not vaporize with the application of heat.

Polyurethane

Polyurethane formulations provide excellent humidity resistance and far greater chemical resistance than acrylic coatings. They require very lengthy cure cycles to achieve full or optimum cure. Removal of polyurethane coatings can be difficult due to their very high resistance to solvents. For single-component coatings, the preferred method of removal is via burn-through for spot repair, while specially formulated strippers enable the user to completely remove the coating from an entire PCB assembly for more wide-ranging rework concerns.

Epoxy

Epoxy coatings are very hard, usually opaque, and good at resisting the effects of moisture and solvents. Epoxy is usually available as a two-part thermosetting mixture that shrinks during curing, leaving a hard, difficult-to-repair film. Epoxy possesses excellent chemical and abrasion resistance but can cause stress on components during thermal extremes. Epoxy is quite easy to apply but nearly impossible to remove without damaging components.

Acrylic

Acrylic conformal coatings are perhaps the most popular of all conformal coating materials due to their ease of application, ease of removal, and forgiving nature. Acrylics dry rapidly, reaching optimum physical properties in minutes; and they are fungus-resistant and provide long pot life. Additionally, acrylics give off little or no heat during cure. This eliminates potential damage to heat-sensitive components. They do not shrink during cure, have good humidity resistance, and exhibit low glass transition temperatures. The material also has a continuous operation range of -65 to +125 C.

Parylene

Parylene is applied at room temperature using chemical vapor deposition (CVD) to apply a very thin, optically clear coating that covers all exposed surfaces. The cost of parylene compared to other, more conventional approaches is much higher, so its application is limited due to the costs involved.

Superhydrophobic

The definition of superhydrophobic coatings is that they have a wetting angle far greater than the 90 degrees typically defined as hydrophobic. These materials can create barriers far more resistant to humidity and condensation than standard conformal coatings. Hydrophobicity tends to be driven by the number and length of the fluorocarbon groups and the concentration of these groups on the surface (Caswell 2015).

For the family of coatings, the key points are similar:

- Assisted chemical vapor deposition (CVD) processes
- Room temperature deposition process
- Low vacuum requirements
- Variety of potential coating materials (with a primary focus on fluorocarbons)

A key differentiation in superhydrophobic coatings is how they break down the monomer before deposition. These coatings can be pinhole-free at 100 nm or less. The primary advantages are that the materials are optically transparent, have RF transparency, and are reworkable, and the masking operation to protect components can be eliminated.

From a DfM perspective, it is vital to assess the Tg of the coating for the operating temperature environment. If the Tg is within the operating range, high stresses can impact solder joints and components due to the change from compressive to tensile stress as the temperature cycle passes through the Tg. Also, PCB cleanliness is paramount and needs to be part of the DfM assessment before the conformal coating operation. The ionic contaminants of greatest concern are primarily anions, especially halides (chlorides and bromides). These are chemically aggressive due to their chemical structure and are very common in electronics manufacturing processes. These anions can decrease pH; few metal ions found in dendrites are soluble at mid to high pH. Cu dendrites require a pH of less than 5 to form. Silver(I) ions are soluble at higher pH, which is why silver forms dendrites so easily.

Proper curing of the conformal coating is also a DfM concern. Curing methods include air, UV, thermal, and moisture-laden atmospheres. The time to cure is a function of the type of coating and the application method, with the objective of a tack-free surface. If using a UV-curable coating, you may need a secondary cure for material not exposed to the UV, and the maximum temperature during curing should be less than 100 C. If thermal curing is used, it may require several hours of air curing to permit outgassing before entering a chamber and must be cured to optimum properties before any other environmental exposure.

5.6.6 Potting

Potting materials behave similarly to conformal coatings as they are also designed to protect electronics from environmental, chemical, mechanical, thermal, and electrical conditions that could damage the product. The selection of the wrong potting for an application could result in damage from the potting due to unwanted stresses or heat. Though there are potting materials made from polyurethane, silicone, and UV-cured acrylic, most potting applications use epoxy compounds due to their nice balance of mechanical, thermal, electrical, and adhesion properties.

One of the most common issues with selecting the right potting material is understanding the thermal requirements. Potting materials are typically selected based on the minimum and maximum temperatures that are acceptable. This does not take ramp times and dwells into consideration; and failure to consider dwell and ramp times can lead to over-specifying the materials. For example, if a material with a 200 C continuous rating is selected, it can withstand a short burst at 250 C during a soldering operation. Ignoring the short dwell time could result in selecting a much more expensive material than is required.

Typically, manufacturers select the potting material with the fastest cure cycle. A risk is that a fast cure can result in a larger exothermic reaction, which can cause damage (at greater than 200 C). Fast cures also have the potential to trap air bubbles, which impacts the material's electrical and mechanical properties. The selection of a one- or two-part material can also have an impact, as the easiest approach may not be the best approach. Also, using more potting material increases the risk associated with the exothermic reaction during curing, especially in thicknesses greater than one-quarter to one-half inch.

Determining whether the cure cycle is performing correctly is a part of a DfM analysis. But what happens if the cycle is wrong? For example, there was a situation where components were being damaged by a breakdown of the potting material. Testing showed that the Tg for the material was -22 C and not -40 C as advertised. The failure was observed, and it was noted that the potting material seemed very viscous rather than fully cured. An analysis of the uncured potting material was performed. It demonstrated that the uncured potting material could carry a current, which resulted in a breakdown failure. The uncured material carried a significant and growing level of current with only 20 V applied.

As with conformal coatings, the cleanliness of the surface is vital in achieving good adhesion of the potting.

5.6.7 Underfill

Underfill materials are typically filled epoxy with a high modulus (10 GPa or higher) and a range of CTE values from 16 to 30 ppm. The addition of an underfill improves product performance in mechanical shock and vibration environments by reducing stress on the interconnects due to substrate bending. Underfilling can also improve thermal cycling robustness by reducing shear stress on the solder joints. It is also used to link the die and substrate to reduce thermal expansion mismatch. Like other materials, if an underfill is used around its glass transition temperature, its CTE typically changes more rapidly than its modulus; this can lead to a large increase in the expansion with a negligible change in the modulus, as shown in Figure 5.19. Changes in the CTE in polymers tend to be driven by changes in the free volume. Changes in modulus tend to be

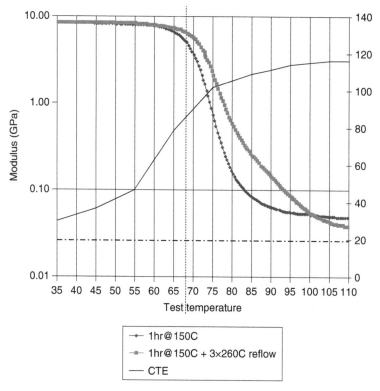

Figure 5.19 CTE and modulus change.

driven by increases in translational and rotational movement of the polymer chains. Increases in CTE tend to initiate before decreases in modulus because lower levels of energy (temperature) are required to increase free volume compared to increases in movement along the polymer chains. As such, understanding the underfill material used is an important element of a DfM assessment (Ebewele 2000).

5.6.8 Cleaning Materials

A DfM analysis must include an assessment of cleaning materials and processes, particularly as the use of bottom-terminated components (BTCs) has dramatically increased. This has caused a demand for cleaning chemistries and processes that can support this type of component.

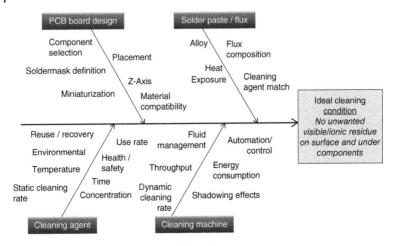

Figure 5.20 Cleaning process considerations.

Other advances include space constraints on the PCB, performance (particularly speed), and the operational environment for the assembly in question. The continuing trend toward miniaturization and more integrated systems with packages like system in a package (SiP) and system on a chip (SoC) have resulted in tighter cleaning requirements. This means the cleaning process must consider all the elements shown in Figure 5.20.

A DfM review should touch on these points to help ensure optimum cleanliness.

5.6.9 Adhesives

Like many of the materials listed in this section, the DfM engineer must have a total understanding of the thermal environment the product may encounter when selecting an adhesive material system. This is an even bigger issue for aerospace, downhole (oil and gas) exploration, and other applications that can require tolerating temperatures up to 300 C. Similarly, some applications must tolerate cryogenic temperatures. Epoxies would typically be the first choice, but newer materials can also handle these temperature extremes.

Picking the right material seems straightforward, but good design practice is not that simple. Published temperature resistance values on data sheets are often inconsistent, primarily because suppliers test adhesives so differently. The Tg issue previously noted for conformal coating, potting, and underfill also applies to adhesives. In the case of adhesives, the Tg is the point where the material goes from a rigid state to one that is much more pliable. This indicates whether an adhesive can handle the application temperature.

5.7   Summary: Implementing DfM

By following the approach detailed in this chapter, a viable DfM process can be performed early in the design cycle, offering a significant improvement to the product and enhancing its ability to achieve its required life expectancy.

References

Binfield, S. (2015a). Flux characterization. DfR Solutions white paper.

Binfield, S. (2015b). How DfR can help you control ionic contamination. DfR Solutions white paper.

Caswell, G. (2013a). Proper stencil design for enhanced manufacturability and reliability of the microsemi 164 pin dual row QFN. DfR Solutions white paper.

Caswell, G. (2013b). Proper stencil design for manufacturability and reliability. DfR Solutions white paper.

Caswell, G. (2014). Conformal coating, why, what, when, and how. DfR Solutions white paper.

Caswell, G. (2015). Superhydrophobic coating. DfR Solutions white paper.

Caswell, G. (2015a). Design for reliability: Part selection. DfR Solutions white paper.

Caswell, G. (2015b). Capacitors: Reliability issues. DfR Solutions.

Caswell, G. (2015c). Design review for reliability improvement. DfR Solutions webinar. https://www.dfrsolutions.com/resources/design-review-for-reliability-improvement.

Caswell, G. (2016). Understanding and mitigating EOS and ESD in electronics. DfR Solutions webinar. https://www.dfrsolutions.com/resources/understanding-and-mitigating-eos-and-esd-in-electronics.

Caswell, G. and Tulkoff, C. (2014). Non-functional pads: Should they stay or should they go? SMTA ICSR.

Ebewele, R.O. (2000). *Polymer Science and Technology*. Boca Raton, FL: CRC Press.

Galyon, G.T. (2005). Annotated tin whisker bibliography and anthology. *IEEE Transactions on Electronics Packaging Manufacturing* 28 (1): 94–122.

Galyon, G.T., Xu, C., Lai, S. et al. (2005). The integrated theory of whisker formation: A stress analysis. In: Proceedings Electronic Components and Technology, 421–428. IEEE.

Gulbrandsen, S., Arnold, J., Caswell, G., and Cartmill, K. (2014). Comparison of aluminum electrolytic capacitor lifetimes using accelerated life testing. IMAPS Symposium.

Han, S., Osterman, M., Meschter, S., and Pecht, M. (2012). Evaluation of effectiveness of conformal coatings as tin whisker mitigation. *Journal of Electronic Materials* 41 (9): 2508–2518.

Hillman, C. (2010). How to control tin whisker risk by implementing appropriate mitigations. IPC Tin Whiskers Conference.

Kadesch, J.S. and Brusse, J. (2001). The continuing dangers of tin whiskers and attempts to control them with conformal coating. NASA EEE Links newsletter.

Kadesch, J.S., Leidecker, H., and Day, J.H. (2000). Effects of conformal coat on tin whisker growth. NASA NEPP.

Lee, B.-Z. and Lee, D.N. (1998). Spontaneous growth mechanism of tin whiskers. *Acta Materialia* 46 (10): 3701–3714.

Masterbond (2018). High temperature resistant adhesives beat the heat. White paper.

Tulkoff, C. (2013). Contamination and cleanliness challenges. IPC Conference on Solder and Reliability Materials, Processes and Tests.

Tulkoff, C. (2014). Design for manufacturing: Challenges and opportunities. APEX Expo.

Woodrow, T.A. and Ledbury, E.A. (2006). Evaluation of conformal coating as a tin whisker mitigation strategy, part II. In: *Proceedings of the SMTA Internal Conference*, Rosemont, IL, 24–28. Citeseer.

6

Design for Sustainability

6.1 Introduction

Design for Sustainability (DfS), also known as Design for Life-Cycle Management, is ultimately about designing to avoid obsolescence issues and ensuring that electronic products are repairable and supportable. The decisions made during the design and development process greatly influence the ultimate life-cycle results and costs experienced by end-users. DfS is a comprehensive design strategy that incorporates:

- Design for Reliability (DfR) +
- Design for Manufacturability (DfM) +
- Design for Repairability and Testability (DfT)
- = Optimized total ownership cost

Planning for obsolescence is a critical element of Design for Life-Cycle Management. Obsolescence management is a key driver that also mitigates the risk of counterfeit parts. The secondary market is frequently the supply chain of last recourse when a part becomes obsolete or has constrained supply. While it is possible to get high quality, authentic parts in the secondary market, it is also possible to get nonconforming, reworked, or counterfeit components. Component obsolescence management enables businesses to anticipate and plan for supplier disruption, end of life parts, aging technologies, and long-life programs. Companies that don't prepare for obsolescence are extremely vulnerable to counterfeit parts in addition to quality and reliability challenges.

Design for Excellence in Electronics Manufacturing, First Edition.
Cheryl Tulkoff and Greg Caswell.
© 2021 John Wiley & Sons Ltd. Published 2021 by John Wiley & Sons Ltd.

6.2 Obsolescence Management

Managing component obsolescence issues is critical to long-term reliability (Tulkoff and Caswell 2012). Companies producing high reliability, long-life products should anticipate and plan for obsolescence by implementing a robust management program that considers:

- Asset security
- Component inspection
- Product genealogy (origins) and condition
- Storage environment
- Data management
- Ensured supply
- Potential quality and reliability issues

Rapid changes in technology, uneconomical production requirements, and extended product life cycles are major causes of component obsolescence. The US Department of Defense (DoD) term for component obsolescence is *diminishing manufacturing sources and material shortages* (DMSMS). DMSMS can impact any phase of the product life cycle from development to post-production. It affects both cost and operations. If left unchecked, DMSMS can cause increased downtime due to part shortages and decreased reliability if alternative solutions are not thoroughly qualified (Tomczykowski 2003). To relate obsolescence to non-DoD products, consider the rapid technology advancements in consumer products such as personal computers and cellular phones over the last decade, and how often they need to be upgraded or replaced.

6.2.1 Obsolescence-Resolution Techniques

Use of Existing Stock
This is the use of obsolete, surplus components currently owned by the product manufacturer or in-stock at a warehouse. Activity considerations include monitoring the possibility of dormant failure rates, preventing improper storage, and handling, and preventing improper application due to lot differences or part suffix differences.

Reclamation of Parts
Reclamation can occur at the system, module, circuit board, or component level. Because of potential reliability problems, the reclamation

of components from circuit boards is not recommended. Even though removal processes have improved recently, the reclamation of components should only be implemented as a last resort or temporary measure until other resolutions can be determined. Reliability, maintainability, and supportability (RMS) considerations include refurbishment of circuit boards, conducting electrical and qualification testing, and vigilance against inducing failures.

Alternate Parts

An alternate part is a part that is equal to or better than the part specified on a BOM. Such parts may be listed in a specification or standard as superseding parts; upgraded or better than original parts; or equivalent or interchangeable parts that are functionally the same, mechanically the same, and of the same quality as the specified parts like parts from a different vendor.

Substitute Parts

A substitute part is a part whose performance is less capable than the part specified on a BOM for one or more reasons (e.g., quality or reliability level, tolerance, parametric, temperature range). RMS considerations include:

- Conducting form, fit, and function testing.
- Performing qualification conformance inspection (QCI) testing.
- Performing system-level or in-circuit evaluations.
- Minimizing the effect on reliability through careful selection, testing, and most importantly, understanding the physics of a failure of the design, and installing it using best manufacturing practices. This assumes that timing and other critical characteristics meet the design operating requirements.

Aftermarket Parts

Aftermarket manufacturers such as Rochester Electronics and Lansdale Semiconductor are authorized by the original equipment manufacturer (OEM) to provide custom assembly of obsolete ICs using existing wafer and die. In some cases, a manufacturer has acquired the photomask sets or the entire assembly process from the OEM. A photomask holds the pattern for each layer of an IC design. The manufacturer is then authorized to produce wafer, cut die, and package the cut die. In some cases, finished goods provided in a catalog may be considered as a substitute part. The RMS considerations for the aftermarket are the same as for substitute except that in-circuit evaluations are always needed due to variations in manufacturing processes.

Lifetime Buys

A lifetime or last-time buy is the purchase of a sufficient quantity of a soon-to-be-obsolete item to meet the projected demands of the supported equipment for its expected operational lifetime. Unfortunately, lifetime buys seldom obtain the correct quantity of parts; either too many or too few components are procured, especially if the expected operational lifetimes are extended. RMS considerations are the same as those for the existing stock resolution technique. A bridge buy – a purchase of limited-quantity components to support near-term requirements until a longer-term solution can be achieved – is a preferred solution.

Emulation

Emulation is a manufacturing process that produces a substitute form, fit, function item for the unobtainable item. Through microcircuit emulation, inventory reduction can be achieved because obsolete items can be replaced with devices that emulate the original and can be manufactured and supplied on demand. RMS considerations for emulation are the same as for the aftermarket.

Circuit Board Redesign

Circuit board redesign may be classified into three basic categories. A Type 1 redesign typically involves adding jumper wires or components. A Type 2 redesign typically involves a board re-layout to change not more than a few components. A Type 2 redesign is also known as a minor redesign. A Type 3 redesign, often called a major redesign, involves designing a DMSMS item out of the system and typically replaces the entire board with a form, fit, and function replacement. Type 3 redesigns are usually a last resort but provide the opportunity to enhance system performance and improve reliability and maintainability. Because substantial non-recurring expenses (NRE) and recurring logistics cost accrue for Type 3 redesigns, they are most appropriate when a sufficient quantity of obsolete components is involved for the same circuit board (rules of thumbs have been discussed as greater than five to ten components). The decision to implement a Type 3 redesign is made by conducting a cost trade-off analysis unique to each circuit board. An exception is made when safety or performance measures are not meeting the requirements. RMS considerations include:

- Awareness of induced failures possible for Type 1 and 2 only
- Awareness of an increased probability of failure for Type 1 and 2 only
- Revising and updating maintenance procedures
- Conducting mandatory qualification and system testing
- Anticipating software rehosting and testing

6.2.1.1 Industry Standards

For companies that proactively manage component availability and obsolescence, the effect of long-term storage is the area of greatest concern. Depending on the technology and storage environment, failure mechanisms to consider include solderability, intermetallics/oxidation, stress-driven diffusive voiding, moisture, Kirkendall voiding, and tin whiskering. Of all these, solderability and wettability remain the number-one challenge.

To create a comprehensive component obsolescence program, it is helpful to review some of the available standards. The most relevant industry standard for storage reliability is the ANSI-GEIA-STD-0003, "Standards for long-term Storage of Electronics." This document was created to provide an industry standard for long-term storage (LTS) of electronic devices by drawing from the best long-term storage practices currently known. In the standard, LTS is defined as any device storage for more than 12 months. However, component storage times are generally much longer. While the standard focuses on unpackaged semiconductors and packaged electronic devices, it may be applied to other components as well. In the standard, electronic devices are defined as any packaged electrical, electronic, electro-mechanical item, or assemblies using these items. Unpackaged semiconductors are semiconductor wafer or die. The standard is intended to ensure that adequate reliability is achievable for devices after long-term storage. It does not replace the need to request data from suppliers or to generate internal data that demonstrates successful storage life for the duration desired by the user. The standard discusses the appropriate storage conditions, containers, and documentation needed for the three levels defined within. Sub-reference standards within the ANSI/GEIA-003 include:

- MIL-PRF-27401, "Propellant Pressurizing Agent, Nitrogen"
- MIL-PRF-81705, "Barrier Materials, Flexible, Electrostatic Protective, Heat-Sealable"
- JESD 625, "Requirements for Handling Electrostatic Discharge-Sensitive (ESDS) Devices"
- JESD-033, "Handling, Packing, Shipping and Use of Moisture/Reflow Sensitive Surface Mount Devices"

Any components or engineered commodities that are expected to be stored for one year or longer must be handled according to the Level 3 requirements in ANSI-GEIA-STD-0003 Standards. The General Reliability Storage Checklist should also be filled out.

Stored materials that will be soldered or soldered to must have solderability assessed according to the Set A requirements of IPC/EIA JEDEC J-STD-002C, "Solderability Tests for Component Leads, Terminations, Lugs, Terminals, and Wires." This standard prescribes test methods, defect definitions, acceptance criteria, and illustrations for assessing the solderability of electronic component leads, terminations, solid wire, stranded wire, lugs, and tabs. After the first year of storage, solderability should be assessed. Solderability should be assessed again before the use of any partial or complete material sets. Certain, but not all, materials may be reworked to enhance solderability.

The MIL-HDBK-338B "Electronic Reliability Design Handbook" provides another viewpoint on the issue. For many reliability predictions, if the equipment is off or non-functioning, the failure rate is assumed to be insignificant, perhaps even zero. Field evidence and experimental data, however, show that assumption to be false. Many types of components can experience failures even when no electrical stresses are applied since other stresses are still present. Some of these other stresses include temperature, acceleration, shock, corrosive gases, and humidity. Variations in all these can be experienced in a long-term storage environment.

For some components, the storage failure rate can be worse than the expected operating failure rate and at lower stress levels.

6.2.1.2 Asset Security

Companies may choose to create an obsolescence and long-term storage program, or they may elect to choose a provider for the service. Regardless of who manages the program, the critical elements to consider are identical. Asset security is vital to protect against loss and theft and simply to keep track of components over the long-term. Any component going into storage must be subjected to some form of incoming inspection process that confirms both authenticity and conformance to quality requirements. Documentation of the product genealogy (origin) and condition is also required, including data for manufacture, transportation, and any prior short-term storage. Any environmental data, lot codes, and date codes should also be tracked. The same component from different lots and date codes may behave very differently after storage. The storage environment should be maintained as per the ANSI-GEIA-STD-0003 standard with active desiccant storage at less than 5% relative humidity (RH) or dry nitrogen storage per MIL-PRF-27401.

Long-term data management systems must also be capable of maintaining and managing individual date and lot codes over time. Parts may be pulled and used at varying times in partial or full lots. And some testing during storage must be performed. This "continuing" data needs to be captured and updated to maintain the integrity of the information. Finally, the obsolescence program must indeed ensure the parts supply needed over the duration needed. Maintaining data appropriately helps give insight as to how well the system is working. Data integrity also helps give an early warning if supplies are depleting or degrading faster than anticipated. This provides the procurement organization more time to respond appropriately and avoid having to purchase from suppliers outside approved distribution channels.

For companies that don't want to manage the logistics of a component obsolescence program, some businesses offer managed supply programs (MSPs). Services provided include purchasing and holding of obsolete components, long-term storage, component contract financing, stock pooling and optional stock holdings, product quality inspection and management, and contract terms up to 20 years. Contract or stock pooling options involve paying a percentage of the cost of the parts over some defined time interval from the manufacturer or MSP provider. Less purchase investment is required by avoiding the upfront cost of completely purchasing all the parts needed. The percentage payment ensures that the parts are stocked and available when needed. This option also results in less inventory cost to a business and reduced risk of losing or damaging stocked parts. Storage space is not needed, and the warranty period doesn't begin until a part is purchased from the pool. With purchased parts, the warranty period starts on the actual date of purchase. This can be a significant advantage if problems arise.

Now that some of the overall elements of an obsolescence program have been reviewed, long-term storage specifics should be considered. Long-term storage presents many challenges. Some issues include the need for physical space and cost and warranty considerations. However, with appropriate care, ICs can be successfully stored long-term at the die/wafer level or as finished goods packaged in plastic or hermetic packaging. Long-term storage technically begins at the one-year mark. In the commercial world, two years is considered long; whereas in the military, space, and avionics fields, 20 years and beyond is quite common.

6.3 Long-Term Storage

Clearly, for long-term programs, some form of storage should be considered. Long-term storage may present problems including locating practical/physical space, mechanical and financial concerns, and risk of counterfeit products. Long-term storage in the commercial sector may be as little as 2 years, while long-term storage in high-reliability applications can be 20 years or longer. With appropriate care, ICs can be reliably stored at the die/wafer level, or as finished goods (packaged) for the long term. Certain industries, however, may be prohibited from using long-term storage by regulations such as the Federal Acquisition Regulations (FAR). FAR often limits procurement to one or two years. Systems manufacturers for projects governed by FAR have rarely funded long-term procurement from their budgets. This may be a situation where considering some of the managed supply options is useful in limiting financial risk and preparing for obsolescence.

Successful storage methodologies for die and wafer include special bagging, environmental controls, and periodic monitoring. Proper storage requires care, cleanliness against particulates and gases, and benign temperatures. Integrated device manufacturers (IDMs) perform die banking, but few distributors do. A controlled atmosphere is created using "dry boxes" with dry nitrogen purged storage or using dry bagged parts and vacuum storage. Oxygen barrier bags designed specifically for long-term storage should be used. Storage at the die/wafer level provides several key advantages. First, the packaging is extremely compact, so it requires little space. The form factor is also very flexible. The die can be assembled into almost any desired package, including packages that may not currently be available.

Hermetic packaging minimizes moisture intrusion, with 20-year storage times considered routine. Common hermetic packages include the metal TO-3 "can" and ceramic and side-brazed packages in dual in-line package (DIP), flat-pack, and pin-grid array (PGA) formats. The packages must be kept dry and in environments low in sulfur, chlorine, and hydrocarbons to preserve the solder finish on the lead frame. Hermetic packages pose both disadvantages and advantages. Their package type cannot be changed, and they are slightly more expensive to store than the die-banking option. Large storage space is required, but it is an easy storage infrastructure with long lifetime.

Misconceptions about plastic packaging are pervasive. Some organizations assume that since parts come from the manufacturer in sealed packaging, they don't require any further special handling or storage; or, since some parts are not rated with moisture sensitivity level (MSL) requirements, they are risk-free and safe to store in any normal factory or warehouse environment. Nothing could be further from the truth. Plastic is hygroscopic: it attracts water molecules from the environment. A plastic package can achieve moisture equilibrium in 4 to 28 days, depending on the molding compound used (TI 2006). Although it can take up to four days for a part to reach moisture equilibrium, it takes longer for actual damage to occur. A "normal" room environment is considered "wet" for plastic ICs. As an example, data from an LAX weather station collected over 30 years shows an annual indoor RH average of +70%! So, plastic packages must be stored in moisture barrier bags (MBBs or dry bags) or in a 10% or less RH environment.

Many people are surprised at the concept of a normal room as wet. But they forget that, in operation, a device is turned on, the die heats up, and the moisture is driven out. Components, however, are not stored in

Table 6.1 Storage Options Summary.

	Die or Wafer	Plastic	Hermetic
Total cost	Low	Medium	High
Form factor flexibility	High	Low	Low
Stability/lifetime	High	Low	High
Storage space	Small	Large	Large
Storage infrastructure	Moderate	Difficult	Easy

a powered-up condition. People also correctly state that water doesn't hurt plastic. However, it's not the plastic that is of concern; it's the water. Water leaches or reacts with materials from the mold compound, elements in the gases in the environment, and other materials deposited on the outside of a package. Water also corrodes and degrades metal pads and wires, which drives device failure.

Some plastic components are also rated as non-moisture-sensitive. However, this rating is for IC to board assembly for reflow solder to prevent heat-induced delamination and popcorning. Contrary to popular belief, it is not a rating for long-term storage.

Wafer, die, and hermetic and plastic packages can be reliably stored for long periods. All must be stored in a clean, dry environment. Plastic finished goods also require periodic monitoring. Having stored components essentially eradicates the problem of locating the end of life (EOL) obsolete parts in the future. Table 6.1 summarizes the storage options.

6.4 Long-Term Reliability Issues

Long-term storage reliability issues to consider include intermetallics and oxidation, stress-driven diffusive voiding, moisture, Kirkendall voiding, and tin whiskering.

Solderability
Concerns with poor solderability driven by oxidation formation can be potentially mitigated using more aggressive flux formulations.

This requires contingency planning for assembly of components after long-term storage, including changing flux chemistries and introducing modified cleaning processes to ensure that these chemistries are effectively removed after soldering. The most critical parameter to control during long-term storage is temperature, since oxide formation can be potentially remedied, while issues like intermetallic formation cannot.

Intermetallic Formation

Intermetallic compounds form when two unlike metals diffuse into one another, creating species materials that are combinations of the two. Intermetallic growth is the result of the diffusion of one material into another via crystal vacancies made available by defects, contamination, impurities, grain boundaries, and mechanical stress. There are several locations within the electronic package where these dissimilar metals are joined. These include die-level interconnects and wire bonds, plating finishes on lead frames, solder joints, flip-chip interconnects, etc. Growth of intermetallics during the storage period may occur and may reduce strength. Growth may also increase the resistance of the interconnect due to the properties of the intermetallic or due to Kirkendall voiding.

Intermetallic layer thickness can be estimated by the following equation:

$$X = (k \times t)^{1/2}$$

where X is the intermetallic layer thickness, t is the time, and k is the rate constant that is calculated by the following:

$$k = Ce^{-KT}$$

where C is the rate constant (nine different ones are listed by Philofsky), e is the activation energy (typically 0.4 to 0.9 eV), K is the Boltzmann constant, and T is the temperature in absolute scale. So, modeling and prediction can be performed for this mechanism.

Stress-Driven Diffusive Voiding

Stress-driven diffusive voiding in on-die interconnects results from a mismatch in the coefficient of thermal expansion (CTE) between the dielectric layers and the metallization itself. Aluminum has a very high CTE ($\approx$27 ppm/degree C), while SiO_2 has a low CTE ($\approx$ 4 ppm/degree C). Since the metal deposition operations during semiconductor manufacturing are performed at elevated temperatures, the metallization contracts as it cools, causing it to be in a tensile state. These tensile stresses relax over time, resulting in small movements (diffusion) of metal atoms. This movement can result in a void or an open interconnect. Compressive stresses from the molding process can also cause the movement of metallization atoms. This can cause thinning of the interconnect, resulting in greater current densities during operation. These current densities may be high enough to cause electromigration, which can also lead to an interconnect open.

Tin Whiskers

The formation of tin whiskers on the surface of tin-plated lead frames may cause a reliability issue. A tin whisker is a single-crystal growth that can occur on tin-plated lead frames. The mechanism for the growth is not clearly understood, but it does appear to be related to compressive stresses in the plating, moisture, and contamination. Whiskering may also be an issue for alloy 42 lead frames with pure tin platings, since large compressive stresses are present due to the CTE mismatch between the alloy 42 and the tin. Tin whiskers can lead to shorting, intermittent errors, and high-frequency issues.

Moisture

Depending on storage time and conditions, parts may be subjected to moisture. This can occur due to the overloading of the desiccant with moisture, failure of storage bags, or improper storage. The presence of moisture can lead to corrosion issues and other failures such as popcorning during subsequent soldering operations.

Printed circuit board (PCB) storage requires some unique storage considerations. Common lead-free (LF) board final finishes include electroless nickel/immersion gold (ENIG), immersion tin (ImSn), immersion silver (ImAg), organic solderability preservative (OSP) and LF hot air solder leveling (HASL). If SnPb HASL plated boards have historically been used, the biggest change will be allowable storage times. Except for ENIG, which many companies avoid because of cost, and perhaps LF HASL, alternative LF platings should be limited to 12 months of storage. Studies of the storage life of LF HASL platings are underway, with some showing two- to three-year storage life or better. Over time, ImSn will form intermetallics (temperature), OSP-coated copper will oxidize (humidity), and ImAg will tarnish (gaseous sulfides). OSP-coated boards may potentially be reworked once or twice, but ImSn and ImAg finishes may not. Once these finishes have degraded, the PCBs must be scrapped.

Kirkendall Voiding

Another potential issue with board plating that involves solder is Kirkendall voiding. Kirkendall voids may also be referred to as champagne voids, since they resemble bubbles in a glass. This issue occurs when voids form at the interface between two dissimilar materials due to differential diffusion. If these voids coalesce, solder joint failure is more likely, especially under mechanical shock/drop conditions. Figure 6.1 shows an example of Kirkendall voids.

Summary of Failure Modes of Stored Electronic Components

Table 6.2 summarizes failure modes that may occur due to long-term component storage.

Figure 6.1
Kirkendall or champagne voids. Source: Ansys-DfR Solutions.

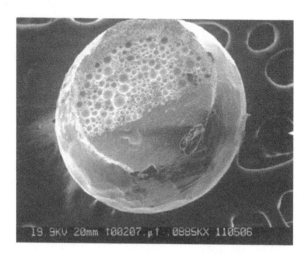

Table 6.2 Failure modes of stored electronic components.

Component	Failure modes
Batteries	Dry batteries have limited shelf life. They become unusable at low temperatures and deteriorate rapidly at temperatures above 35 C. The output of storage batteries drops as low as 10% at very low temperatures.
Capacitors	Moisture permeates solid dielectrics and increases losses, which may lead to breakdown. Moisture on plates of an air capacitor changes the capacitance.
	Moisture causes changes in inductance and loss in Q. Moisture swells phenolic forms.
Coils	Wax coverings soften at high temperatures.
Connectors	Corrosion causes poor electrical contact and seizure of mating members. Moisture causes shorting at the ends.
Relays & Solenoids	Corrosion of metal parts causes malfunctioning. Dust and sand damage the contacts.
	Fungi grow on coils.
Resistors	The values of composition-type fixed resistors drift, and these resistors are not suitable above 85 C. Enameled and cement-coated resistors have small pinholes that bleed moisture, accounting for eventual breakdown. Precision wire-wound fixed resistors fail rapidly when exposed to high humidities and to temperatures at about 125 C.
Diodes, transistors, and microcircuits	Plastic-encapsulated devices offer poor hermetic seal, resulting in shorts or opens caused by chemical corrosion or moisture.
Motors, blowers, and dynamotors	Plastic parts swell and rupture, and metal parts corrode. Moisture absorption and fungus growth on coils. Sealed bearings are subject to failure.
Plugs, jacks, and dial-lamp sockets	Corrosion and dirt produce high-resistance contacts. Plastic insulation absorbs moisture.
Switches	Metal parts corrode, and plastic and wafer warp due to moisture absorption.
Transformers	Windings corrode, causing shorts or open circuits.

6.5 Counterfeit Prevention and Detection Strategies

When component obsolescence isn't planned for, the secondary market is often the supply chain of last recourse. While it is possible to get high-quality, genuine parts, it is also possible to get nonconforming, reworked, or counterfeit components – and it is increasingly difficult to differentiate genuine parts from their counterfeit equivalents. Historically, the secondary market provided a mechanism for finding parts in short supply or at a reduced cost. Today, high-reliability system manufacturers are less willing to risk contamination of their supply chain with potentially substandard parts to save a few dollars on the cost of a part. The proliferation of counterfeit components has led to a contraction of the secondary market and an increase in the cost of parts in the marketplace (Caswell 2010).

In some ways, counterfeit components that do not function are the least critical issue. While they represent a source of financial loss to the system manufacturer, their failure is at least identifiable during the system manufacturing process. Components that are substandard or reworked are much more difficult to manage. These parts may have been reclaimed from other systems. In some cases, they include scrapped components that failed a manufacturer's testing or quality control program. This group of components represents a significant risk because they may function at the time of manufacture but may fail in the field. In high-reliability systems, not only is the cost of a field failure significantly higher than the cost for commercial systems, but the failure may put lives at risk.

As an example, obsolete parts are a key driver of counterfeits entering the DoD supply chain. Between 50 and 80% of suspect counterfeit parts were obsolete when they were reported. The Defense Logistics Agency (DLA) checks and applies DNA authentication technology to every microcircuit it procures (over 80,000 annually). A DNA mark enables rapid screening of the microcircuit throughout the supply chain and retrieval of pedigree information anytime throughout its life (Jones 2016).

Unfortunately, counterfeit parts are no longer an emerging threat. They're real, and they're here to stay.

The Impact of Counterfeits

Counterfeits impact cost, reputation, security, and failure. The risk and impact vary based on the type of counterfeit and operational environment. Costs to the semiconductor manufacturer include both visible and hidden costs. Visible costs include revenue loss, device-level anti-counterfeiting measures, and personnel measures. Personnel costs are driven by training, tracking, inspecting, testing, and documentation. Documentation must be created for the procedures, analysis, and control activities. Hidden costs may include avoiding potentially lower-cost manufacturing companies or regions of the world, negative impact on company reputation and brand, and, most importantly, loss of credibility to buyers.

Equipment producer costs include the need for training, limiting potential sources, and implementing purchasing controls. Incoming product inspection requires time, tracking, and specialized test equipment. Extra control and testing of product sourced from non-approved or infrequently used sources is necessary. All this effort results in lost manufacturing time and output while resourcing devices if counterfeits are found or delivered. However, it costs much more in rework, scrap, and liability if counterfeits get into the manufacturing process.

End component user costs include the need to provide additional servicing and spares due to premature equipment failure and equipment downtime. Counterfeits can lead to defects or bugs that may cause equipment to malfunction. Legal expenses may be incurred due to downtime, damage, failure to perform, injury, or loss. Finally, military and national security may be threatened. Military aircraft have been repeat victims of counterfeits ranging from ICs to counterfeit memory to use of reworked parts.

Counterfeit Definitions

A counterfeit may be:

- Any part or component that is not what it is represented to be
- Duplication of another manufacturer's product
- Manufacturer's product that failed test, inspection, or burn-in or was an engineering sample
- Fake or "up-revision": Empty package, incomplete functionality, lesser specs

- Salvaged or reworked parts (used)
- Trojans: Introduce malware or tracking

Other relevant definitions include:

New

In most cases, a product that has never been sold is legally considered new.

Used

Products that have been sold to external customers and used by them for any length of time.

Refurbished

Used or returned product; product repaired with new parts, re-tested, and warranted; product re-tested and warranted because it couldn't be determined if it had been used; or new product built with a combination of new and used parts (due to obsolescence), tested, and warranted.

Demo

Products used for demonstration purposes (trade shows, etc.). These products have seen gentle use, are re-tested before sale, are sold at a discount, and come with a limited warranty.

For the DoD, *counterfeit electronic part* means any unlawful or unauthorized reproduction, substitution, or alteration that has been knowingly mismarked, misidentified, or otherwise misrepresented to be an authentic, unmodified electronic part from the original manufacturer, or a source with the express written authority of the original manufacturer or current design activity, including an authorized aftermarket manufacturer. Unlawful or unauthorized substitution includes used electronic parts represented as new, or the false identification of grade, serial number, lot number, date code, or performance characteristics.

Figure 6.2 shows a summary of the definitions.

Obstacles to a Successful Anti-Counterfeit Program

Some companies fail to create an anti-counterfeit program due to complacency. Because they haven't identified a problem yet, they believe that "it can't happen here." Identifying an internal owner of the program can

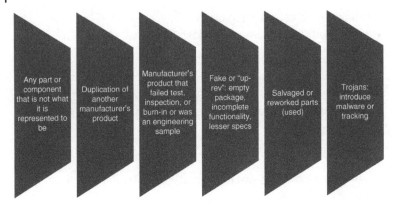

Figure 6.2   Counterfeit definitions.

also be a challenge. The procurement organization takes the lead in most companies. Not understanding or underestimating the threat is also common. Frequently heard counterfeiting myths include:

- Only expensive parts are counterfeited.
- Buying from an "American" company prevents problems.
- The parts are the latest design and can't be faked.
- The distributor provides a certificate of conformance (C of C).
- The receiving inspection or test process will catch fakes.

Ironically, the Federal Acquisition Regulation (FAR)/ Defense (DFAR) rules may impair the ability to prevent counterfeits. FAR/DFAR rules say to use the lowest bidder and need to meet Small Disadvantaged Business quotas. There's also the belief that "If I restrict where I buy from, I will pay more money." The supply chain may also come up with creative excuses ranging from cost to liability. One common lament is that other customers don't have a requirement. Another frequent refrain is that the activity is just too expensive. Obviously, affordability is a key concern. Counterfeit mitigations must pass the cost vs. outcome test.

Cost-Effective Counterfeit Protection Approach
A cost-effective anti-counterfeit approach begins with research. With a strong understanding, an organization can then carefully select sources, implement purchasing best practices, and close the loop by performing lot-validation testing.

Anti-Counterfeit Best Practices

- Contact the original component manufacturer (OCM) and franchise distributors first.
- Confirm product status; is the product near or at end of life (EOL)?
- Check for authorized aftermarket sources.
- Determine if/when the product went EOL.
- Determine EOL production lot date codes.
- Try to locate a known-good device.
- Check with industry sources like ERAI or the Government-Industry Data Exchange Program (GIDEP) concerning counterfeit activities for the product.
- Clearly communicate objectives and needs to suppliers.
- Create a company policy on counterfeits.
- Look for actual product in stock, not just availability.
- Ask if the product is traceable back to the OCM.
- Choose sources wisely.
  - Don't use cost as the sole or primary criterion for purchase.
  - Use industry sources to evaluate potential suppliers.
  - Ask for accreditations, certifications and memberships like ISO-9001, AS9100, IDEA-1010, ERAI, or IDEA membership.
  - Audit sources.
  - Purchase only from suppliers that have implemented and are committed to counterfeit avoidance practices.
  - Understand warranty and payment terms.
  - Not all brokers and independent distributors are the same. Many have minimal quality systems and no electrostatic discharge (ESD), storage, or handling controls. Some have no experience with semiconductors.
  - The internet has created virtual companies with no traceability back to the original OCM and limited or no product warranty or support.
  - Create a list of trusted suppliers, and use them.
- Buying obsolete product requires more diligence, not less.

Before buying, ask for photos and evidence for the product. Objective evidence consists of images of the component front and back, date codes, EOL dates, and markings. Compare date codes and EOL dates provides to manufacturer data, and compare marking with known-good products. Understand the product warranty – anything less than 30 days is a red flag. If the supplier does not have confidence in the product, why would

you? Do the research upfront to save money later. Don't buy COD; nego-
tiate terms, or use escrow services. This practice allows time for authenticity checks and testing to be performed before paying.

Define a Test Plan Based on the Acceptable Level of Risk   Key considerations when creating a risk-based test plan include:

- What is your application? Commercial? Military?
- Define your test plan to meet risk mitigation needs.
- Put requirements in writing and be as detailed as possible.
- Require documented reports of all validation tests.
- Require a copy of the lot information detailing all tests performed and
  results: die ID photos, X-ray photos, etc.

Choose a Reputable Test Laboratory   Select a test laboratory based on its
accreditations and capabilities, not based on price alone. Provide any test
requirements to the laboratory, and compare its test plan against your
risk-mitigation needs. Understand exactly what you are paying for and
who will be performing the testing on your material.

Primary Industry Standards

SAE standard, AS5553   "Counterfeit Electronic Parts; Avoidance, Detection, Mitigation, and Disposition," was developed in 2009 and aims to
standardize requirements, practices, and methods for counterfeit parts
risk mitigation. This document is intended for use in aviation, space,
defense, and other high-performance/reliability electronic equipment
applications. This standard is recommended for use by all contracting
organizations that procure electronic parts, whether such parts are
procured directly or integrated into electronic assemblies or equipment.
The requirements of this standard are generic and intended to be
applied/flowed down to all organizations that procure electronic parts,
regardless of type, size, and product provided.

IDEA-STD-1010   "Acceptability of Electronic Components," was the first
quality standard for the visual inspection of electronic components and
was designed as a technical resource to serve the electronic component
industry regarding the detection of substandard and counterfeit components.

IACC The International Anti-Counterfeiting Coalition Inc. (IACC) is a Washington, D.C.-based non-profit organization devoted solely to combating product counterfeiting and piracy. IACC's mission is to combat counterfeiting and piracy by promoting laws, regulations, directives, and relationships designed to render the theft of intellectual property undesirable and unprofitable.

Defense Federal Acquisition Regulation Supplement (DFARS) (US Government) Requirements for Anti-Counterfeit Programs

- Training of personnel
- Processes to abolish counterfeit parts
- Processes to trace parts
- Using suppliers that are the original manufacturers
- Reporting counterfeit parts
- Procedures to identify suspect counterfeit parts.
- Designing, operating, and maintaining systems to detect and avoid counterfeit parts
- Flow-down of counterfeit detection and avoidance requirements to subcontractors
- Processes for keeping informed of current counterfeiting trends
- Processes for screening GIDEP reports
- Establishing a plan to control obsolete electronic parts
- Inspection and testing of electronic parts
- All material identified as "suspect counterfeit" will be disposed of by a prequalified reclamation facility. All disposed of products will be verified by certificates of destruction (CODs).

Counterfeit Sources

There are four common sources of counterfeit parts: inside jobs, competitors, used components, and fraudulent sources. An inside job is characterized by parts that failed a production test and should have been disposed of but rather are packaged and labeled as good parts of the same type. Depending on the reason the part failed the production test, an inside job counterfeit may operate in benign environments, but it may not function in the more demanding environments described in the specification sheet. The most straightforward way to identify an inside job type of counterfeit is to perform rigorous testing of the part in all environments and functions listed on the specification sheet.

Another type of counterfeit is a part from company B being misrepresented as a part from company A. This may or may not lead to field failures. A common method of identifying these types of counterfeits is to scrutinize the packaging. Many large companies have complex labeling schemes that are difficult to replicate.

A used counterfeit is a part that is used but represented as new. The parts are commonly desoldered from failed circuit boards. They find their way back into the supply chain and have an unknown history and unknown life expectancy. Additionally, the process of desoldering may cause additional damage. A careful inspection of the leads and package for damage or wear and tear should identify a used part. This counterfeiting problem has been exacerbated by waste electrical and electronic equipment (WEEE) as more devices are being salvaged and resold as new to avoid dealing with the disposal of the materials.

A fraudulent counterfeit is a part that is packaged to appear as original and new. Fraudulent parts may contain either an empty package or the wrong chip. These types of counterfeits will fail immediately, as the functional component is phony and serves no purpose other than to appear authentic. These parts only need to appear authentic long enough to be purchased, since they will be detected when they fail to function. This is the most likely type of counterfeit found from goods purchased in the spot market, where the vendor may disappear after the sale is complete and payment is delivered.

The impact of this activity can have a profound impact on the user. Table 6.3 illustrates the various risk-at-failure scenarios as a function of the sources of components.

The potential risk increases with each supplier category. Similarly, the potential risk increases with using sources of supply that are less trustworthy.

Table 6.3 Counterfeit risk and cost.

Source of Components	Probability of Counterfeit	Risk of failure
Component manufacturer	0.02%	$10K to $100K
Licensed distributor	0.20%	$100K to $1M
Broker (known)	2.00%	$1M to $10M
Broker (unknown)	20.00%	$10M to $100M

In the growing world of lead-free, there is now concern over counterfeit RoHS notification. As more companies are required to transition to lead-free, there is an increased likelihood that fraudulent test reports will accompany a component. Major manufacturers that post such data are unlikely to take part in fraudulent activity, but small vendors may be cornered into falsifying RoHS compliance data when faced with the possibility of lost sales due to RoHS noncompliance.

Performing a risk/reward assessment is necessary to determine how much should be spent on methods to identify and detect counterfeit components. Some companies take a very simplistic black or white approach to this issue. A more comprehensive approach is recommended. Using a simple ROI for a typical product of 5–10 to 1, spend between $5000 and $10,000 if the cost to fail is $100,000. More should be spent if the sources of supply require the use of brokers to meet delivery schedules. The number of different part types procured can also add a dimension to this issue as the higher quantity of line items on BOMs may drive the increased use of brokers in a tight economy. Higher component throughput also increases risk, and the amount of time and effort to mitigate counterfeit components should be increased accordingly.

Counterfeit Reporting

For counterfeit or suspect counterfeits, do not return parts to the supplier! Segregate, quarantine, and recall parts as appropriate. Report the counterfeit to the stakeholders, the GIDEP and ERAI databases, and the authorities. GIDEP is a cooperative activity between government and industry that seeks to reduce/eliminate expenditures of resources by sharing technical information. It maintains a suspect counterfeit database at www.gidep.org. Suspect counterfeit reports are special non-conformances under the Failure Experience Data (FED). Manufacturers and GIDEP participants can submit suspect counterfeit data to the GIDEP Operations Center electronically, but a company does not have to be a member of GIDEP to submit data to the program.

In addition to hosting a suspect-counterfeit database, companies like ERAI specialize in counterfeit mitigation if help is needed. ERAI is an information services organization that monitors, investigates, and reports issues affecting the global hi-tech electronics supply chain. It provides tools to mitigate risks on substandard parts, counterfeit parts, vendors, and even customers.

Counterfeit Detection

Most counterfeits are finally detected when they fail during use. However, a visual inspection could raise the suspicion of forgery long before the component fails. Differences in manufacturing specifications such as molding die locations, ink precision or durability, font, or date or lot code standards can indicate possible counterfeit parts. Although knowledge of exterior labeling and marking standards helps detect a counterfeit, the body of knowledge required to become an expert on all standards is extensive. Some counterfeiters have such advanced techniques that the counterfeit marking may be of higher quality (more durable, vibrant, or sharp) than the original. Overall, visual inspection is an important tool in counterfeit detection, but by no means is it the only or best way to identify forgery.

There are more conclusive methods than visual testing to determine the authenticity of components, and many of them diagnose non-destructively. Infrared imaging or SQUID (superconducting quantum interference device) microscopy can be used to identify and monitor the active parts of an IC with questionable components on it. If a known authentic IC is compared to a suspicious one, these imaging techniques will show any current location or size differences. X-ray inspection is the next step and will show any size or configuration abnormalities in die, wire bonds, or bond pads. Hiding abnormalities within a circuit board or component is a popular method of counterfeit since extra effort must be taken to identify them.

Counterfeit Analysis Techniques Counterfeit analysis always starts with a non-destructive evaluation (NDE) that is designed to obtain maximum information with minimal risk of damaging or destroying physical evidence. Always use the simplest tools first. Non-destructive techniques include:

- Visual inspection
- Electrical characterization
- Acoustic microscopy
- X-ray microscopy
- X-ray fluorescence

Frequently performed evaluation techniques include external visual inspection (microscopy), a surface swab or marking permanency test,

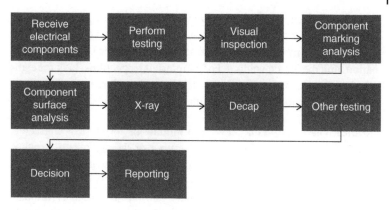

Figure 6.3 Basic validation process flow.

X-ray inspection, and decapsulation and cross-sectioning combined with scanning electron microscopy (SEM) and energy dispersive X-ray spectroscopy (EDX) analysis. These methods are discussed in more detail in Chapter 7.

Basic Validation Process Flow

First, remember that not all suspect counterfeits are counterfeits. Companies have been known to make component changes without issuing a product change notice (PCN). Although these changes may not impair form, fit, or function, they can result in differences in internal visual appearance. Figure 6.3 illustrates a component-validation process flow that can be followed to evaluate whether a component is authentic.

Testing Techniques

Surface Swab Test Blacktopping is a technique where a counterfeiter takes components, usually discarded, and sands them to remove the original component marking. The components are then resurfaced and printed with fake markings. Poor blacktopping techniques can be identified by using an acetone swab test. The acetone dissolves the blacktop and reveals the underlying component. However, the methods used to blacktop sanded parts have improved, so the acetone test is no longer considered a reliable indicator.

X-ray X-ray analysis may be used to verify the die and wire bonding pattern of an IC. If a known-good part is available, a direct comparison can be made. An X-ray can also uncover defects such as die bonding delamination, and die attach voiding. For counterfeit inspections, look for extra wire bonds, incorrect configuration of bond pads, different die sizes, extra heat capacity (large paddle), and different lead frames.

Electrical Characterization There are limitations to physical characteristic testing. Parts may be re-manufactured or factory rejects in what appears to be authentic packaging. Or there may be no conclusive identification information on the die once decapsulated. For these reasons, some electrical analysis is helpful in determining authenticity. Although full datasheet testing is highly recommended for high-reliability applications, a curve trace performed on each lead is an effective way to begin an electrical examination of the parts. A known-good (reference) component is required for comparison.

Higher Risk Non-Destructive Inspection Techniques Non-destructive techniques can induce some damage and should be used with care. Since acoustic microscopy requires immersion in fluid, moisture-sensitive components are vulnerable. X-ray microscopy at elevated exposure levels, especially when combined with laminography, may induce damage in sensitive components such as EEPROMs. In thermal imaging, the temperature rise can damage or mask fragile conductive filaments via joule heating, thermal runaway, or evaporation of retained moisture.

Mechanical Robustness Thermal cycling, shock, or drop may be used to verify that a component is not counterfeit. Results are compared to previously evaluated parts.

Decapsulation and Cross-Sectioning After performing all the relevant non-destructive tests, IC components may be opened (de-packaged) for die verification. Decapsulation and cross-sectioning is a form of destructive physical analysis (DPA) that enables direct inspection of circuits and identification of missing or extra circuitry, layers, and interconnects. It is an invaluable method for determining the authenticity of difficult-to-resolve parts. Since it is possible on individual components, the method does not guarantee all the parts in the lot are the same, but it is typically cheaper to perform than electrical testing.

Testing and Reporting Recommendations

Choose analysis labs carefully, since there is wide variation between labs. Clearly identify the information desired from the analysis. There are no industry standards or accreditation for laboratory counterfeit analysis, and the counterfeit landscape changes rapidly.

Recommended Information:

- Investigator and lab name
- Datasheet information
- Assessment checklist
- Analysis and test descriptions and tools
- Electrical test results
- Overall comments
- Results
- Summary

Recommended Image List – As Needed:

- Packaging images
- X-ray images
- XRF images
- EDS images
- Acoustic microscopy images
- Overall images of part
- Lead condition images
- Solvent testing images
- Internal images
- Corner radius images

Examples of Device-Level Countermeasures

Encryption on a chip is often cited as one of the answers to counterfeiting woes, but it is difficult to use for small components. Intel announced a process that allows the random-number generator, which is the basis for encryption, to be made with the same semiconducting material and at the same feature size now used for modern processors. Smart marking involves an ink containing DNA or other biologic markers. The ink marking is visible under infrared light. RF tagging is also used for package authentication; tagging makes it extremely difficult to counterfeit. The data on RFID tags acts as an electronic security marker to automatically authenticate genuine product packaging.

Manufacturers' countermeasures include complex markings on the package, use of ultra-violet ink in package marking, and symbols or identifiers etched in the die. These techniques make it difficult to alter parts and enhance traceability. One of the latest techniques involves etch-depth analysis: the use of digital light processing (DLP) to rapidly scan surfaces for highly accurate depth measurements. Simply compare the data to either the manufacturer's spec or a known-good part.

Mitigating Effects of Counterfeits

Do a full performance screen that tests to or beyond the specifications. Design the test to detect fraudulent parts empty packages or incorrect functionality. A performance screen will not detect parts with latent defects or those with additional circuitry.

To prevent counterfeits, do comprehensive audits of distributors. Review their quality system. Perform a complete risk assessment of the program, and develop an inspection and test plan that addresses the risks.

Anatomy of a Counterfeit Process: Steps Involved in Repackaging a Die

Step 1: Decapsulation Decapsulation, often a very dangerous, dirty, and manual process, is the first step in the overall process to repackage a die. Red fuming nitric acid or sulfuric acid is heated to 140 C. Then, the component is repeatedly dropped into the solution to remove the plastic material covering the die. To maximize profit, counterfeiters are under extreme pressure to perform, which increases the chance for accidents with the oxidizing acids and their fumes. The chemicals used are aggressive toward both the die and bonding pads.

Step 2: Bond Wire Removal (Plucking) With fine tweezers and steady hands, a counterfeit operator gets under the wire and pulls the wires in the direction opposite the ball bond attached to the pad. Plastic tweezers are preferred since they tend to cause less damage. The most difficult part of this process is breaking the wire at the top of bonds of a decapped part without damaging the original ball bond.

Step 3: Chip Removal (Extraction)   To extract the die from the package and lead frame, the counterfeiter uses a hot plate at 100 C to loosen the adhesive holding the die to the lead frame. Then, a utility blade and tweezers are used to pry the die off the lead frame or package mounting.

Step 4: Removing Adhesive Backing   After the die is extracted, adhesive is left on the back of the die. An operator sands the back to remove the adhesive; this process leaves small abrasions that could later impact the die position.

Step 5: Bonding in New Package   The new wire bond is placed on top of the old ball, which provides some evidence of what has occurred.

Conclusions
Counterfeits are a real and growing threat. The complex global supply chain provides multiple entry points for counterfeits. The risk needs to be managed through a decision matrix where probability, risk, and volumes are considered. This helps ensure clear boundaries and guidelines for mitigation practices. Anti-counterfeiting costs need not be prohibitive and should be chosen as a function of the risk-assessment process.

6.6   Supplier Selection

This section addresses general needs to consider when selecting suppliers along with specific needs when selecting and auditing PCB fabricators and contract manufacturers. When selecting suppliers, using a data-centric approach is recommended. A data-centric approach uses measured behavior to determine the risk of the supplier and risk of the part or product. This approach arises from the realization that "test-to-spec" information is insufficient for high-reliability applications. Companies increasingly use internally developed specifications or modified industry specifications. For this approach to be successful, seek the answers to these questions:

- What are the product margins? This is equivalent to HALT testing.
- Is the supplier in control?
- Can defects be controlled and captured?

- Is the part reliable enough?
- Is the part intrinsically defect-free?
- Has the part been tested to failure?
- Which suppliers and what parts will be subjected to data-centric requirement(s)? All parts? Critical parts? Engineered (non-commodity) parts?

Next, determine the supplier qualification requirements. They may include:

- On-site audits.
- An internally derived questionnaire (self-audit).
- Information on the quality management system used (ISO9000, QS9000, etc.).
- Is the supplier certified to similar or more rigorous customer or industry (medical, avionics, etc.) standards?
- Desired assessment frequency: one time, quarterly, semi-annually, or annually.
- Quantitative scorecard.

Be flexible in the approach used. Identify a company's strengths and weaknesses. If audits are used, keep the audit focused on a minimum set of relevant questions to improve the responsiveness and receptiveness. Assess not just what the company does, but how information is used to control and improve performance. Is there truly preventative action and continuous improvement? What is the company's reliability capability? How far back does the feedback loop go? Is there an ongoing assessment of quality control measures? Determine if processes are sufficiently controlled.

When evaluating quality measures, ask for statistical process control (SPC) data: Cpk or Ppk of critical to quality (CTQs) parameters based on failure modes and effects analysis (FMEA) and average outgoing quality (AOQ) data. For reliability, ask for results from independent reliability testing, review qualification test data, and test-to-failure (HALT) data. For screening data, ask for percent fallout.

SPC involves using a sample of data to estimate the percentage of product that meets specification limits. *Process capability* is the range of values expected from a process over an extended time period. Cpu is the capability based on the upper specification limit (USL), and Cpl is the capability based on the lower specification limit (LSL). Pp measures the process spread vs. the specification spread. A process is capable if the range of expected values fall within the specification limits. SPC is based on the capability indices Cp and Cpk, where:

- Cp = (USL − LSL)/6 * standard deviation
- Cpl = (Mean − LSL)/3 * standard deviation
- Cpu = (USL − mean)/3 * standard deviation
- Cpk = Min (Cpl, Cpu)
- A desirable Cpk is equal to or greater than 1.67.

SPC is often based on a minimum of 25–30 samples and represents a snapshot in time that is focused on the future.

An engineered part is anything the supplier is building to your specifications. This includes PCBs, circuit assemblies, housings, etc. Since these are not commodity products, design plays an even larger role in the product's quality/ reliability.

So, what does a data-centric supplier evaluation look like? A PCB fabricator should have all chemical processes under statistical process control (SPC) with all CTQs identified for additional SPC. The CTQs for board manufacturers include:

- Cleanliness and contamination
- Spacing and soldermask thickness (high voltage)
- Plating yield strength and ductility (high aspect ratio plated through-holes)
- Capability (PCQR2)

The following additional information is required or supplied:

- Certificate of compliance (CoCs) on incoming materials
- Glass transition temperature (Tg) – laminate/prepreg
- Hollow fiber concentration – laminate/prepreg
- Copper foil adhesion – laminate
- Specification of soldermask material
- Specification on packaging material

A data-centric assessment of suppliers is critical for reliability assurance since a large percentage of field returns can be linked to component issues. It also optimizes resources by acquiring data primarily on critical components.

6.6.1 Selecting a Printed Circuit Board Fabricator

The objective of the PCB supplier selection process is to choose PCB suppliers that are capable of reliably providing the prototype and production quantities of PCBs required at competitive prices. It is also vital to choose suppliers capable of providing the front-end engineering support of the designs. This support could take the form of assistance in material selection, Design for Manufacturability (DfM), plating, and surface finishes (Tulkoff and Hillman 2012).

Initially, each PCB fabricator should be required to provide an assessment of its capabilities and sources of verification data for evaluation. The best way to accomplish this task is to request that the fabricator fill out the Printed Board Manufacturer's Qualification Profile found in IPC-1710. A completed profile contains the supplier's site capability, a list of its processing and test equipment, technology specifics, details of its quality program, manufacturing history, company data (financial and technical), and data verification sources. As an element of this process, request a copy of the supplier's quality manual, typical work instructions, and SPC data.

Once the data has been reviewed, the supplier should be visited to ensure that it can produce the products.

Having multiple approved circuit board suppliers is recommended; and not all PCB suppliers specialize in the same type of board production. For example, some facilities specialize in layer counts from 2 to 8, others 8 to 16, and yet others to >16 layers. Having suppliers in each of those zones ensures the best pricing and delivery because the supplier is in the sweet spot for fabrication. The trick in selecting a fabricator is to match its business model, manufacturing tooling and processes, materials inventory, and technical expertise to the PCB types that are to be made by it. Implied in this statement is the need to accurately understand the characteristics of each PCB design and to understand what characteristics of each supplier best match its needs.

An assessment of specific board fabrication requirements is also needed. For example, don't select a supplier to fabricate PCBs with 4 mil lines and spacing if the supplier only does this a small percentage of the time. The risk is too great. Recognize that some suppliers can be approved for low-technology products and still be reliable, but not approved for higher technology designs.

As another example, a casual conversation with the CAM operator may reveal a lot about their role in the organization. One manufacturer may have the CAM operator doing a complete design rule check (DRC) on the supplied data and modifying it with venting, thieving, tear-dropping, and compensation to get the highest yield possible; while another merely sends the data out to a third party to generate master film, using the data as is. Most companies would prefer the former.

Incorporate test coupons into circuit board layouts. These coupons are then used for cross-sectional analysis to ensure process stability, quality, and reliability. The PCB fabricator must have the internal ability to cross-section and examine the coupons.

From a general perspective, verify the financial stability of the supplier, obtain references to confirm the quality of the supplier's product, and ensure that the supplier performs value-added services to ensure the manufacturability of the product.

Electronics quality and reliability are highly dependent on the capabilities of the manufacturer. Manufacturing issues are one of the top reasons that companies fail to meet warranty expectations. These problems can result in severe financial pain and loss of market share. What a surprising number of engineers and managers fail to realize is that focusing on processes addresses only part of the issue. Supplier selection also plays a critical role in the success or failure of the final product. Designing PCBs today is more difficult than ever due to significantly increased density, higher lead-free process temperature requirements, and associated changes required in manufacturing. Many changes have taken place throughout the entire supply chain regarding the use of hazardous materials and the requirements for recycling. The RoHS and REACH directives have caused many suppliers to the industry to change their materials and processes. Everyone designing or producing electronics has been affected.

PCBs should always be considered a critical commodity. Without stringent controls in place for PCB supplier selection, qualification, and management, long-term product quality and reliability are neither achievable nor sustainable. Here, we discuss some best practices and recommendations for improving your PCB supply chain.

First, create a PCB commodity team with at least one representative each from design, manufacturing, purchasing, and quality/reliability. The team needs to meet regularly to discuss new products and technology requirements in the development pipeline. These meetings can take place through conference calls or via onsite visits. Pricing, delivery, and quality performance issues with existing PCB suppliers should be reviewed. The team is also responsible for identifying new suppliers and creating supplier selection and monitoring criteria.

Second, establish initial PCB supplier selection criteria. The criteria should be unique to your business and products, but some general selection criteria include:

- Time in business
- Revenue
- Employee turnover
- Training program
- Certified to standards you require (IPC, MIL-Spec, ISO, etc.)
- Capable of producing the technology you need as part of the supplier's mainstream capabilities
- Doesn't build in process niches where suppliers claim capability, but little volume is built there
- Has quality and problem-solving methodologies in place
- Has a technology roadmap
- Has a continuous improvement program in place

Ideally, suppliers demonstrate long-term stability from both quality and business perspectives. Multiple factory locations and risk-mitigation programs can further protect against catastrophic political and weather-related events. After this initial screening, visit the PCB suppliers that seem to be appropriate matches. Site visits to the actual fabrication facility by personnel knowledgeable in PCB fabrication techniques are crucial. The site visit is the best opportunity to review process controls, quality monitoring and analytical techniques, storage and handling practices, and conformance to generally acceptable manufacturing practices. It is also the best way to meet and establish relationships with the people responsible for manufacturing your product.

After choosing some potential PCB suppliers, the true qualification work begins. Qualifying a PCB supplier is always a two-step process. The first step is to qualify both a design and the PCB manufacturer through rigorous product and test-vehicle-based evaluations. PCBs built for qualification should be evaluated to the standards you require by both the PCB supplier and your company (or an independent lab). The second qualification step is to initiate ongoing testing to monitor outgoing quality through a combination of test vehicle and lot qualification.

How do you verify that a PCB supplier can meet your reliability requirements? There are currently six common methods for testing and qualifying PCBs:

- Modeling and simulation
- Cross-sectioning combined with solder float/shock tests
- Thermal shock testing (also thermal cycling)
- Printed board process capability, quality, and relative reliability (PCQR2)
- Interconnect stress testing (IST)
- Highly accelerated thermal shock (HATS)

Very early in the design cycle, software simulation and modeling can be performed. As an example, ANSYS has implemented IPC-TR-579 into an automated design analysis software called Sherlock to permit rapid assessment of basic PCB robustness. IPC-TR-579 documents a "Round Robin Reliability Evaluation of Small Diameter (less than 20 mil) Plated Through Holes in PWBs." This activity was initiated by IPC and published in 1988. The objectives were to confirm sufficient reliability, benchmark different test procedures, evaluate the influence of PTH design and plating, and develop a model. Some advantages of the IPC-TR-579 model are that it is analytical, validated through testing, and comments on the relative influence of design and material parameters. However, the validation data is over 18 years old. The model is unable to assess complex geometries, and it is difficult to assess the effect of multiple temperature cycles. The model also uses simplified assumptions (linear stress-strain above yield point). Finally, the model does not account for the effect of via fill and does not consider other failure modes such as knee cracking and wall-to-pad separation.

To successfully model a PCB design, the first step is to define the environment. Test, field, or both environments can be modeled. The second step is to upload the actual design information. Thermal maps can also be imported, if appropriate. The third step is to select the laminate and prepreg material. With this information, the stackup and copper percentage are automatically identified. Then, PTH fatigue analysis can be performed. Another method is to qualify the design and manufacturer through the PCQR2 database and program. The program consists of a coupon design, a test standard, and a database of participating suppliers. The coupon has an 18 × 24 in. layout with 352 1 × 1 in. test modules. It can consist of two to 24 layers (rigid, rigid-flex) and requires builds of three panels and three non-consecutive lots with six times simulated assembly and HATS cycling.

An advantage of the PCQR2 program is that it is available in an industry-standard (IPC-9151) format. It is easy to follow and provides real data for understanding PCB supplier capabilities. It also provides a comparison to the rest of the industry using an anonymous database. However, it is an IPC-approved monopoly and can be expensive, not including panel costs, per supplier and facility evaluated. Interconnect stress testing (IST) is the overwhelming favorite of high-reliability organizations for lot-based qualification testing. IST uses small (1 × 4 in.) coupons that can fit along the edge of a product panel. Testing is automated, is widely used, and can drive barrel fatigue and post-separation failures. The large number of holes (up to 300) and continuous resistance monitoring make it far superior to cross-sectioning. And, in most cases, IST is a more cost-effective option. There are some potential issues to be aware of, however. The IST coupon design is critical. Preconditioning, test frequency, maximum temperature, and failure requirements must all be specified. Test frequency can range from every lot to every month to every quarter. Maximum temperatures of 130 C, 150 C, and 175 C are most commonly used. And, different markets and organizations specify different times to failure, with 300, 500, and 1000 cycles most common.

Next, if your product line ranges from simple to highly complex, consider tiering (low, middle, high) suppliers. Tiering permits strategic supplier selection to maximize cost savings and quality against your product design requirements. Match supplier qualifications to the complexity of the product. Some criteria for tiering suppliers include:

- Finest line width.
- Finest conductor spacing.
- Smallest drilled hole and via.
- Impedance control.
- Specialty laminates or construction needed (PTFE, flex, mixed technologies).
- Buried resistance or capacitance.
- Use of HDI, microvias, blind or buried vias.

- Minimize the use of suppliers that outsource critical areas of construction. Again, do not exist in the margins of any supplier's process capabilities.

Relationship management is another important element of success. Ideally, you are a partner with your PCB suppliers. This is especially critical if you have low volumes, low spend, or high technology and reliability requirements. Good partnering practices include:

- Routine conference calls between your PCB commodity team and each PCB supplier's equivalent team.
- Quarterly business reviews (QBR) that review spend, quality, and performance metrics, and include "state of the business" updates. Business updates can address any impending changes, such as factory expansion, move, or relocation; critical staffing changes; new equipment/capability installation, etc. The sharing is done on both sides. Share any data that would help strengthen the business relationship: business growth, new product and quoting opportunities, etc. At least once per year, the QBRs should be site meetings that alternate between your site and the supplier factory. The factory supplier site visit can double as the annual onsite visit and audit that you should be performing.
- Ongoing supplier "lunch and learns" or technical presentations held at your facility or via webinar. Suppliers routinely provide education regarding their processes and capabilities. They can educate your technical community on PCB design for manufacturing, quality, reliability and cost factors. They can also provide information on pitfalls, defects, and newly available technology. This activity is usually performed free of charge, and suppliers may even provide free lunch to encourage attendance. It also provides the opportunity to educate them on your needs. Supplier scorecards should exist and be used on a rolling quarterly and yearly basis. Typical scorecards include both objective and subjective metrics. Objective metrics include on-time delivery, cost, PPM defect rates, and quality excursions that require root cause corrective actions. Subjective measures include communication issues (speed, accuracy, responsiveness to quotes) and cost-savings suggestions. And, there should be a detailed discussion of any recalls, notifications, or scrap events exceeding a certain threshold.

- Make sure that any supplier has a robust continual quality improvement program in place. At a minimum, consider monitoring and reviewing the top three PCB factory defects, process control and improvement plans for those top defects, yield and scrap reporting for your products, and feedback on general issues facing the PCB industry.

The foundation of a reliable product is a reliable PCB. Having a comprehensive strategy for selecting and qualifying PCB suppliers ensures the foundation is strong.

6.6.2 Auditing a Printed Circuit Board Fabricator

A key recommendation from Caswell (2016) concerned the need for an onsite audit of PCB suppliers. This recommendation arises because there is no accepted industry standard methodology for qualifying PCB suppliers. Standards exist for lot-based PCB testing and acceptance within the IPC 6010 series, but sourcing follows the "as agreed upon between user and supplier" (AABUS) approach; however, onsite audits remain the most robust approach (Tulkoff and Hillman 2013).

For an audit to be successful, it must be performed by people knowledgeable in the processes involved in fabricating a PCB. More than 180 individual steps are required in manufacturing a typical PCB. The processes are complex and chemistry-intensive and require tight process controls to guarantee acceptable results. An audit encompasses all levels and types of PCB fabrication from generic, multilayer rigid boards to complex rigid-flex configurations. Audits can be performed to assess the viability of a new supplier, to ensure supplier compliance with customer requirements, or to resolve a problem that has resulted in product failure. Before performing an audit, the information outlined in the preceding sections should be requested and reviewed. A formal report documenting the audit scope, contents, and results should be generated for all audits. The audit report should also identify the individuals who participated in the audit. As mentioned before, people with process knowledge must perform the audit. If a company lacks appropriate resources, it can partner with companies that provide PCB audit experts as a service.

Materials and Process Audit

This section provides insight into the assessment of the process operations observed during a typical audit of a PCB fabrication facility.

Planning Engineering

The initial step in most PCB fabrication facility processes involves the planning engineering function. This is where an engineer verifies the customer data, the quality assurance codes, specifications, and any other special requirements. Commonly referenced industry standards include:

- IPC-1710 Printed Board Manufacturer's Qualification Profile
- IPC-2221 Generic Standard on Printed Board Design
- IPC-2222 Sectional Standard on Rigid Organic Printed Boards
- IPC-6011 Generic Performance Specification for Printed Boards
- IPC-6012 Qualification and Performance Specification for Rigid PCBs
- IPC-A-600 Acceptability of Printed Boards
- IPC-9151 Printed Board Process, Capability, Quality, and Relative Reliability Benchmark

Numerous other IPC and industry standards are available for both processes and test methods. If IPC standards are used, the customer must also specify the appropriate class. The IPC classes are defined as follows:

- Class 1: General Electronic Products (consumer products)
- Class 2: Dedicated Service Products (uninterrupted service is desired)
- Class 3: High Reliability (continued performance or performance on demand is critical)

The planning engineer normally constructs the in-house traveler and identifies the materials to be used for fabrication, the stackup needed to meet electrical requirements, and any impedance controls. Then, the lay-up sheet identifying the copper weights per layer, core thickness, and laminate type is created. A net-list verification and review of the customer statement of work (SOW), engineering change notices (ECNs), IPC-6012 requirements, panel size, and special codes are also performed. If required, the planning engineer also incorporates HATS, IST, or other test coupons into the lay-up. Finally, the planning engineering's work initiates the computer-aided manufacturing/design (CAM/CAD) and document control activities. Verification of all these functions occurs in the first stage a detailed audit.

CAM/CAD

The CAM activity usually establishes a customer part number and drawing for the project, creates a job number, links the input data from the customer, sets up the layers using internal naming structures, and performs any edits needed to match internal processes with the customer supplied data. CAM also checks the customer-supplied data against the drill files and sets the lines and spaces necessary for each layer. A Design for Manufacturability (DfM) review and design rule check (DRC) of Gerber data for functional integrity and manufacturability are performed. Finally, the CAM operation generates a panelized layout for the project, including the inclusion of specified test coupons. All these actions must be verified during an audit.

Documentation and Configuration Control

Documentation packages are then prepared for each production run and are maintained for a defined time, typically five years, unless the customer requires a different retention period. All standard operating procedures (SOPs) should be readily available for review in this area.

Certification and Training

An auditor should verify the internal certification and training programs in place for employees to ensure that they are all qualified to manufacture the customer's products to the appropriate specifications.

Change Management

The audit should also determine how changes in equipment, processes, and materials are implemented, documented, and communicated, both internally and externally. Even supposedly minor changes can have a substantial impact on fabrication process results.

Fabrication Flow

The following sections describe a typical PCB fabrication flow and recommended assessment approach.

Incoming Receiving The audit should check that the materials received are verified against the list of materials called for in the project. The thickness of the cores, copper weight, laminate material, and layer identification are ascertained at this operation. Verify that the materials selected meet the procurement requirements by picking a random lot for assessment. How prepreg is handled and transferred is a critical operation.

Prepreg should be stored in a temperature and humidity-controlled area at less than 23 C and less than 50% RH. Prepreg should not be folded, and any opened packages should be resealed. If the storage temperature is significantly below room temperature, the prepreg should be allowed to acclimate to ambient conditions before use. Prepreg should only be handled by the edges and with gloves. Typically, prepreg should be moved into the fabrication process on a first in first out (FIFO) basis to properly control storage times. IPC 1601, "Printed Board Handling and Storage Guidelines," documents best-handling practices throughout fabrication.

Inner Layer Processes In the inner layer imaging process, the PCB laminate is coated with photoresist, a photosensitive polymer dry film material, and a positive image is transferred to it. In dry film development, imaged panels are placed in a conveyorized chamber of potassium carbonate, and the unexposed dry film resist is chemically removed. This process results in a pattern formed in the resist on top of the copper surface (a positive image). Next, the developed internal layer panels are placed in a copper etchant chamber. In the developing process, the exposed copper is removed from the panel. The dry film resist covering or protecting the circuitry pattern is then stripped off.

The auditor should examine how the cores are cleaned and how the wet laboratory monitors samples. All processes analyzed should have process controls in place, checks for verification, planned maintenance, and calibration controls that are defined by the panel size. Phototool storage temperature and humidity levels should also be evaluated. The tools should be used and stored under conditions like the actual manufacturing conditions used to make them. The tools are moisture sensitive and grow or shrink due to changes in the storage conditions. Next, audit the imaging process used to produce the artwork and the process used to perform the exposure operation to ensure that the setup and artwork match the panel structure.

Inner Layer Automated Optical Inspection (AOI) and Visual Inspection
The audit must examine the automated optical inspection (AOI) tools used and how they are linked to the CAD data. The AOI equipment is programmed using Gerber/CAD data to inspect the etched copper panels. Every panel is placed in the machine and inspected for defects as

small as .001 in. in diameter. All identified defect locations are verified by operators using visual inspection under magnification. Some type of data collection must be used to record defects found during the inspection of line widths and spacings that are listed on the traveler. All data should be recorded, including information on the type of defect and its location. A layer inspection is audited next. This should include the method used to monitor the materials being transferred into the operation. Good PCB fabricators perform transfers via an airlock to minimize oxidation.

Inner Layer Preparation   The bond film operation is next in the process and is usually fully automated. If necessary, the materials being prepared for lamination are baked, and information is logged into a data collection system. Sets of lay-up materials are generated and are presented to the lamination presses as completed stackups or "books." In lamination, the etched and inspected internal layers are sequenced and registered on large steel plates. Prepreg or B-stage material consisting of fiberglass and a semi-cured resin system is placed between the etched layers. The plates are then inserted into a heated press to bond the layers. The process engineer establishes the recipes for the presses by linking the parameters to the specific job number. The engineer selects the correct recipe and logs it for verification purposes. A controller monitors the temperature, pressure, and vacuum values to specified levels. The laminated boards then go through a cleaning operation to remove excess material from the edges. The thickness of the stackup is verified using a laser thickness measurement system. All data should be recorded. Next, each panel is typically serialized. Test coupons are cut out for microsectioning to confirm that inner copper layers are in the correct sequence and that the correct layers of prepreg have been used. If everything has been processed to specification, the process engineer allows the stackups to proceed to the drilling operation.

Drilling   Before the drilling operation, the stackups are X-ray inspected to set fiducials to estimate the center point for setting the tooling holes. Drilling is normally accomplished using automated drilling systems. The number of drill hits should be monitored through a drill list that defines replacement requirements based on size and speed. Drill bits are wearout items and need to be replaced per the directions on the traveler or specifications. An X-ray inspection system performs reference detection and measurement of the drill coordinates. It also calculates the optimized drill location and determines compensation factors for tooling holes. The laminated panels are registered to the drill machine. A drill program extracted from the Gerber data is uploaded to the drill controller. Drill bits are automatically selected, and the panels are drilled at the programmed locations. X-ray inspection is again performed to verify the accuracy of internal layer alignment and to verify proper hole-pad positioning.

Drill bit manufacturers typically provide PCB manufacturers with recommendations on key process parameters, including:

- Speeds and feeds
- Stackup guidelines (number of PCBs of a given thickness that can be stacked during drilling)
- Entry and exit material
- Number of drilling operations before repointing
- Number of repoints and sharpening frequency

There is no one "right" answer for process parameters. A PCB manufacturer may buy a more expensive drill bit but repoint more often. The key to success is verification and control with a compliance and reward system. Did the PCB manufacturer perform its own design of experiments (DoE) to understand and verify guidelines from the drill bit manufacturer? Manufacturers need the ability to capture the influence of high glass content, heavy copper, fill particles, etc. based on the designs built. Are incoming and resharpened bits subjected to automatic inspection? Are vacuum gauges alarmed and monitored? Is statistical process control (SPC) of drill runout checked before production? Is the PCB manufacturer confirming employee compliance with defined drilling parameters? Are bonuses in line with process parameters where compensation cannot be increased by exceeding process recommendations?

Hole Preparation and Hole Metallization   Hole preparation to accomplish deburring can be performed via several methods, including a brushing technique or a plasma cleaning operation. The audit should verify that all holes are completely clear of debris. Desmear removes resin and drilling debris from the hole wall. Etchback is a controlled removal of resin to a specific depth to expose additional internal layer connection surfaces. Etchback is typically used in high-reliability applications. Cross-sections are performed to verify numerous parameters including hole wall quality, desmear or etchback, plating thickness, and dielectrics. Physical cross-sections of the finished product or digital images are supplied to the customer per their specification. Ideally, each customer should discuss and specify desired cross-section locations and sampling plans. Many organizations understandably focus on the most challenging structures but lose sight of the variation that occurs across the diagonal of a panel or varying hole sizes.

Copper Plating, External Layer Imaging, and Developing The electroless plating process deposits a thin layer of copper from an autocatalytic plating solution without the application of electrical current. Audit all the electroless processing tanks in the production area. At the control console, the operator should select from a list of materials and cycle times for each project and enter that data into a log to confirm the correct selection. The process engineer should be the only person who can change a recipe. Operators should only be able to select a recipe from the menu on the control computer. The sole action required by the operator at this point is to load and unload the panels since all other functions are typically automated. At external layer imaging, the drilled panels are coated with photoresist, and a negative image is transferred onto the dry film surface. The exposed panels are placed in the developer, and the unexposed dry film resist is removed. The result is a pattern of exposed copper. This negative image is then ready for copper plating. Additional copper is added to the foil circuit pattern and drilled holes by an electrolytic copper plating process. A tank is electrically charged with current to activate the electroplating process. The copper plating line also includes several cleaning/etching baths, rinses, and acid copper baths. All must be continuously controlled and monitored.

Like drilling, plating chemicals manufacturers provide PCB manufacturers with guidance on process parameters and equipment. Many will even provide a "turn-key" installation. However, this can result in a lack of onsite knowledge if PCB manufacturers do not perform their own DoE since there is a large variation in plating chemistries, processes, and equipment. Some common plating process variations are as follows:

- Chemistry: Primarily sulfuric acid copper, but several proprietary additives (brighteners, levelers, etc.) may also be used

- Process: Primarily flash, strike, and then panel followed by pattern
  - Thick boards (e.g. 180 mil) with high aspect ratio (12:1). PTV can require additional plating steps (up to three to four total)
- Equipment: Primarily vertical but increasing interest in horizontal conveyorized
  - DC or pulse reversed
  - Soluble vs. insoluble anodes

In some respects, PCB plating is overly complicated since every supplier can have a unique solution. Again, the key to success is verification and control, with a compliance and reward system. Did the PCB manufacturer perform its DoE to understand and verify guidelines from the plating manufacturer? Did the PCB manufacturer capture the influence of power and ground planes, beginning and end of anode life, etc.? Is incoming chemistry tested? How often is the plating line chemistry analyzed? Is it based on time or production volume? Is SPC conducted on actual measurements, or is averaging performed? Is plating thickness immediately checked through eddy current measurements? How often are plating properties evaluated? Is the PCB manufacturer confirming employee compliance with defined plating parameters?

The ANSI/IPC-A-600 standard requires an average plating thickness of 20 μm, with isolated areas allowed to reach 15 μm. Insufficient plating thickness is caused by insufficient current or time in the copper plating bath or poor throwing power. When insufficient plating thickness is observed throughout the PTV, instead of just at the center, the root cause is more likely insufficient current or time in the plating bath.

Copper Plating Analysis PCB fabricators have several pattern plate processes that can be utilized. They include:

- Nickel only
- Fused Sn/Pb
- Hot air solder leveling (HASL)
- Electroless nickel / immersion gold (ENIG)
- Electroless nickel / electroless palladium/ immersion gold (ENEPIG)
- Nickel with hard or soft gold

All the process parameters for each of these variations should be monitored. TrueChem software is highly recommended for this purpose.

Plating Thickness and Material   Plating is considered to be the number one driver for PTV barrel fatigue (Barker and Dasgupta 1993). In a classic engineering conflict, there is a trade-off where gaining better properties (greater thickness, higher plating strength, greater elongation) usually requires a longer time in the plating bath and longer time in the plating bath reduces throughput. This makes PCBs more expensive to fabricate. PCB fabricators, already struggling in a low-margin business, try to balance these conflicting requirements. The key plating parameters are thickness, strength, and elongation (ductility). Plating thickness specifications tend to range from 0.8 mil (20 microns) to 1.0 mil (25 microns) to 1.2 mil (30 microns). Since plating rates have increased substantially due to new formulations to fill microvias, plating thickness can be less of an issue than in the past. Now, PCB manufacturers rarely have issues with thin plating. Some PCB manufacturers, depending on the design and production volume, plate the PTV almost closed. PCB manufacturers also experiment with plating rates to increase ductility, so times to failure are longer even if the plating thickness is not ideal.

Plating Mechanical Properties   Industry standards, such as IPC-6012C, require an ultimate tensile strength greater than or equal to 36,000 psi (250 MPa) and an elongation greater than or equal to 12%. PCB manu-facturers tend to be very familiar with the test requirements specified by their larger or higher reliability customers, so routine plating conditions are adjusted to meet those requirements.

Post-Plating Measurements of Plating Properties   One of the challenges of root-cause analysis of PTV cracking is the inability to directly measure the strength or ductility of the plating. Hardness and grain size mea-surements are potential substitutes. Hardness data generated by Mike Sosnowski of EMC showed a 100% correlation with cross-section results between two populations. Boards with cracks had high hardness; boards without cracks had low hardness.

Resist Strip, Outer Layer Etch, Tin Strip   The plasma etching process for etchback or desmear is the next process operation. The operator should log the film type, exposure level, age of the bath, and chart for day/day settings. To maintain control, etching is normally based on the starting copper weight vs. the speed of the process. It is also at this point in the process that controlled impedances in the board are tested and verified. The plated panels are removed from the copper bath and placed in a tin plate bath, temporarily covering the copper circuitry and creating an etch resist. The dry film is stripped, and the exposed copper is removed by the etching chemistry. The tin plate is chemically stripped. The etched external layers are laser inspected using AOI. All defective areas on the panel surface are then operator-inspected under magnification before solder-mask coating. After these processes are complete, copper thickness and hole sizes are verified, and the panels receive a 100% visual inspection.

Soldermask Processes Resist is typically applied in a clean-room environment per the requirements of the traveler, which should also specify hold time and laminate type. This material is then photo-processed. The operator selects the resist height and logs the data into the collection system or tally sheet and shop order. A laser direct imaging (LDI) process for this function is recommended. Most soldermask processes use a liquid photo imageable (LPI) material, with a 100% inspection of the images after processing. Dry or wet mask processes, if used, should also be audited. In LPI soldermask processing, the panel surface is coated with a polymer and partially cured. The soldermask image is applied to the panel. The unexposed soldermask is then removed in a developer. This exposes the holes and surface mount pads.

Legend/Silk Screening The legend/silk screening application is another highly automated process. The material must be properly applied and cured. A first-article inspection is performed here to ensure complete, correct marking, including date code and serial number. Legend ink should not cover solderable surfaces. Verifying controls for these processes is part of the audit.

Final Finish Commonly available final surface finishes for PCBs include HASL or LF HASL, ENIG, ENEPIG, ImAg, and OSP. The surface finish influences the process yield, amount of rework, field failure rate, ability to test, scrap rate, and cost of a PCB.

The selection of a surface finish should be a holistic approach that considers all important aspects of the assembly. Companies can be led astray by selecting the lowest-cost surface finish, only to find that the eventual total cost is much higher. The process and HATS or IST coupons are removed after this operation and are sent to a cross-section laboratory for analysis or chambers for further testing.

PCB Cleaning   Contamination is believed to be one of the primary drivers of field issues in electronics today. It induces corrosion and ECM. Several industry trends have converged to drive an increased risk of contamination failures. The decreased spacing of the components and circuitry used in electronics increases the impact of any contamination. The increased heat load per unit volume also drives higher airflow, which brings more particulates and gasses into contact with electronics. This has resulted in increased corrosion rates seen in the industry. To minimize this risk, PCBs must exit the fabrication processes as clean as possible. Cleanliness levels must be carefully measured and monitored. Unpopulated PCB cleanliness guidelines and standards are available from IPC. They include:

- IPC-5701: Users Guide for Cleanliness of Unpopulated Printed Boards
- IPC-5702: Guidelines for OEMs in Determining Acceptable Levels of Cleanliness of Unpopulated Printed Boards
- IPC-5703: Guidelines for Printed Board Fabricators in Determining Acceptable Levels of Cleanliness of Unpopulated Printed Boards
- IPC-5704: Cleanliness Requirements for Unpopulated Printed Boards

Verify that the latest cleanliness standards are in use at the facility. Ion Chromatography (IC) is the gold standard method for measuring ionic cleanliness, but very few PCB manufacturers qualify lots based on IC results. Most facilities use IC to baseline resistivity of solvent extract (ROSE), omegameter, or ionograph (R/O/I) results. The recommended best practice is to perform a lot qualification with R/O/I and periodically recalibrate with IC at some specified interval (every week, month, or quarter).

Electrical Testing   Flying probe and fixture or bed of nails probe testing are basic elements of PCB testing. Electrical measurements can be made at different levels using a sampling plan. At the electrical test, a test netlist, usually per IPC-D-356, is uploaded into the tester. Each board is placed on the tester and electrically tested for continuity and resistance. How tested PCBs are handled and segregated at test is critical.

Due to the manual loading nature of the testing process, it can be easy to accidentally co-mingle passed and failed PCBs. Look closely at how tested material is identified and segregated. Auditors should also validate how the procedures for test yield are monitored and how aberrant or low yielding lots are handled.

Final Inspection and End Item Data Package Certified inspectors should perform the final inspections of the circuit boards using stereo-zoom microscopes. On-site training should be provided to these inspectors for all specifications and standards required for the PCBs manufactured at the facility. Not all product parameters require a 100% inspection. It is vital to establish an agreed-upon sampling plan that considers product complexity and reliability needs. IPC-A- 600, "Acceptability of Printed Boards," is widely used for final inspection, but customers may require other standards or custom requirements. If defects are identified, they should be marked for touch-up, if allowed, and then sent to a final inspection after rework. Coupons, cross-sections, and full documentation should be prepared for delivery with the completed product as required by the customer. IPC-QL-653A, "Certification of Facilities that Inspect/Test Printed Boards, Components and Materials," may also be used to verify inspection and test areas. At final inspection, quality control personnel visually inspect 100% of the finished product and verify the product per fabrication drawing requirements for dimensional properties, board size, finished hole sizes, and other special customer specifications. If there are discrepancies, customer specifications always take precedence over referenced industry standards.

IPC-A-600 defines three levels for each inspected characteristic. They are:

- Target: Depicts the desired condition. May not be necessary to ensure reliability.
- Acceptable: The condition depicted, while not necessarily perfect, will maintain the integrity and reliability of the board in its service environment.
- Nonconforming: The condition depicted may be insufficient to ensure the reliability of the board in its service environment. Considered unacceptable for at least one class of product but may be acceptable for other classes as specified by the acceptance criteria.

Characteristics are divided into two basic groups: externally visible or internally observable. Externally visible conditions are features or imperfections that can be seen and evaluated on or from the exterior surface of the board. These characteristics are candidates for visual inspection. Internally observable conditions are features or imperfections that require cross-sectioning of the specimen for detection and evaluation.

Laboratory Analysis: Wet and Cross-Section

As mentioned before, PCB fabrication involves complex, chemistry intensive processes that require tight process controls. The laboratory must validate that bath chemistries, processes, and product results meet all specifications. Requirements can originate from the customer, industry standards, or equipment or chemistry suppliers. All requirements should be documented and continuously monitored using SPC. One recommended software tool is TrueChem, which is designed specifically to control and automate the management of chemistries, coatings, and wet processes.

The IPC provides a complete, free suite of test methods under IPC-TM-650. Many of these are used in PCB fabrication. The methods encompass the following areas:

- Section 1.0: Reporting and Measurement Analysis Methods
- Section 2.1: Visual Test Methods
- Section 2.2: Dimensional Test Methods
- Section 2.3: Chemical Test Methods
- Section 2.4: Mechanical Test Methods
- Section 2.5: Electrical Test Methods
- Section 2.6: Environmental Test Methods

Cross-sections for analysis of the processes should be performed at numerous steps throughout the process. A full laboratory report should be included with the customer's deliverables. These reports should be routinely read and verified by the customer for compliance with specifications. It is not uncommon to find reports showing conformance to IPC Class 2 requirements when Class 3 requirements have been specified. The customer can also specify whether digital images or actual cross-section samples should be provided with the report.

Common PCB Defects

A basic understanding of common PCB defects is helpful. This knowledge can be used by an organization to monitor supplier performance over time. The following descriptions and images illustrate defects that can be seen in cross-sections or microscopy (Tulkoff et al.2013).

Plating Voids *Plating void* is a generic term used to describe voids present in and around the plated through-via (PTV) wall. Plating voids can cause large stress concentrations, resulting in crack initiation. The location of the voids can provide crucial information in identifying the defective process. Voids can be found around the glass bundles, in the area of the resin, at the inner layer interconnects (also known as *wedge voids*), and at the center or edges of the PTV.

Plating voids are areas where the copper fails to deposit or plates more slowly. Voids arise because of the presence of air bubbles caused by supersaturation; supersaturation tends to occur due to air entering into the filter system. A rise in the temperature of the plating bath can also create plating voids by reducing the gas solubility. This results in the nucleation of air bubbles within the plating bath. In the drilling process, the formation of glass voids can be initiated by tearing out a bundle of fiber strands, which creates sites for the packing of debris. The number of voids is inversely related to the temperature, which is directly related to drilling speed. Higher speeds correspond to fewer glass voids. Excessive vibration of the drill bit damages the resin, which is then invaded by the solvent/sensitizer solution. The permanganate bath then preferentially etches this already loosened resin, forming a wedge void. In the desmear process, overaggressive swell due to excessive pH or temperature will often lead to wedge voids. If plated copper is absent on glass fibers tangential to the drilled hole, check alkalinity and activity of the regenerator of the permanganate bath. The presence of permanganate residues, often an indicator of non-optimized neutralizer (insufficient amount, inadequate agitation, high pH, low temperatures), can lead to glass and resin voids.

In the electroless plating process, insufficient conditioner, acceleration, or catalyzation will result in poor or no copper coverage over the glass fibers. A sluggish electroless bath will result in glass and resin voids. Bubbles in the plating bath due to non-optimized bump and vibration agitation result in symmetrical voids primarily in the center of the PTV.

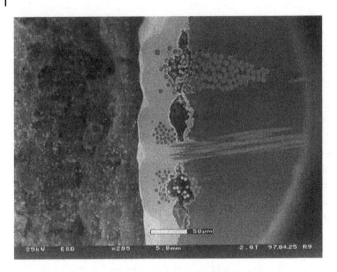

Figure 6.4 Plating voids. Source: Ansys-DfR Solutions.

Dust, lint, or oil films create voids that are less regular in form. In the electrolytic plating process, bubbles in the plating path due to air entering the filter system or excessive temperature rise in the plating bath will result in symmetrical voids primarily in the center of the PTV.

Figure 6.4 shows plating voids.

Glass Fiber Protrusion　Glass fiber protrusion into PTV walls affects PTV plating thickness and can contribute to PTV cracking. Glass fiber protrusion may occur due to process control variabilities during hole drilling, hole preparation, or application of flash copper. Glass fiber protrusion is allowed by IPC guidelines only if the minimum plating thickness is met. Figure 6.5 shows glass fiber protrusions.

Plating Folds　Plating folds create detrimental stress concentrations. Rough drilling or improper hole preparation can lead to plating folds. Rough drilling may be caused by poor laminate material, worn drill bits, or an out-of-control drilling process. Improper hole preparation can occur due to excessive removal of epoxy resin caused by an incomplete cure of the resin system or due to a preparation process, like desmear or etchback, that is not optimized. Figure 6.6 shows plating folds.

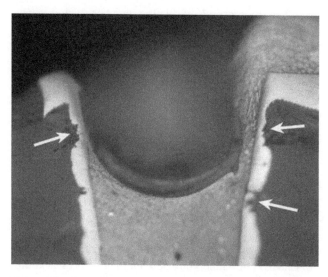

Figure 6.5 Glass fiber protrusion. Source: Ansys-DfR Solutions.

Figure 6.6 Plating folds. Source: Ansys-DfR Solutions.

Plating Nodules The root causes of plating nodules include poor drilling, particles in solution, a solution temperature that is out of range, and excess brightener levels. The presence of plating nodules can be detrimental to high reliability as they create highly stressed areas in the plating wall and can reduce lifetime under temperature cycling. The ANSI/IPC-A-600 standard states that nodules are acceptable if the hole diameter is above the minimum specified. Figure 6.7 shows plating nodules.

Etch Pits Etch pits occur due to insufficient tin resist deposition or improper outer layer etching process and rework. They cause large stress concentrations locally, increasing the likelihood of crack initiation. Large etch pits can result in an electrical open. In the tin resist process, an insufficient or non-uniform thickness can result due to improper bath chemistry, nonuniform agitation, or the presence of inorganic or organic contamination. Check bath temperature and additives. Check acid dip and adjust within 5 to 10%. Verify proper rinsing after the acid dip.

In the etching process, pits can be caused by an overly aggressive process. Materials to check include:

- Ammonium chloride: Check for etchant pH higher than normal and specific gravity lower than normal.

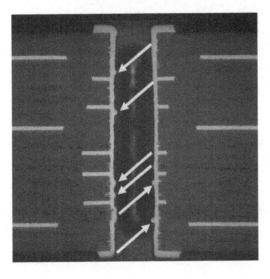

Figure 6.7 Plating nodules. Source: Ansys-DfR Solutions.

- Cupric chloride (CuCl$_2$): Check HCl content for higher than normal (0.1 N) and specific gravity for lower than normal (30–32 Baume at 54 C)
- Peroxide sulfuric: Check peroxide level for higher than normal (spray method, sulfuric acid, 17–20%; hydrogen peroxide, 10–20%) and for bath temperature above range (55 C for spray).

In the cleaning process, cleaning time must be optimized to the PTV diameter and etchant solution used. Very small holes rely on capillary action as the primary rinse mechanism. This is a slower process. Warm solutions rinse faster than cold ones; more concentrated solutions require a longer rinse time. Closely spaced panels are more difficult to rinse, and alkaline solutions take longer to rinse than acidic ones. The design also influences the formation of pits. Poor design practices can result in trapping of etchant residues. Examples of weak practices include the use of single side tented vias or torn or insufficient tenting. Etch pits tend to be located toward the surface of the PTV. Questions to ask regarding etch pits are as follows:

- How does the supplier ensure tin resist thickness?
- How does the supplier monitor etchant chemistry?
- Does the supplier track etchant chemistry using SPC?
- How does the supplier optimize the cleaning process?
- How does the supplier track rework?
- Does the supplier specify the maximum amount of rework allowed?

Figure 6.8 shows etch pits.

Shipping

A requirements document should be used to summarize yield figures, start dates and date codes, along with special notes per customer requirements. At shipping, PCBs are vacuum-sealed and packaged to ensure product integrity and to protect against moisture and contamination. Packaging should incorporate MBBs, desiccant, and HICs. Bar codes and labels are applied per customer specifications. Certificates of conformance, test certificates, RoHS certificates, and any other customer-requested documentation are provided for shipment. Transportation is usually based on customer requirements. As noted before, IPC 1601, "Printed Board Handling and Storage Guidelines," documents best practices.

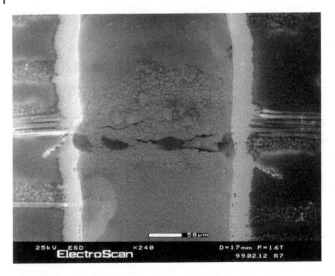

Figure 6.8 Etch pits. Source: Ansys-DfR Solutions.

Final Documentation Review

Audit a typical documentation package for completeness and accuracy. This stage can also be used to review and address any unique or special processes not covered elsewhere.

Conclusion

The audit process described addresses a typical PCB fabrication process. Not all facilities have all the operations described here. Regardless, the flow in the audit should encompass the unique processes to ensure compliance. The foundation of a reliable product is a reliable PCB. Having a comprehensive strategy for selecting and qualifying PCB suppliers ensures that the foundation is strong. Performing effective on-site audits is a critical component of that strategy.

6.6.2.1 Selecting a Contract Manufacturer

Similarly, the contract manufacturer (CM) selected should have the ability to produce assemblies with the defined parts mix, the defined processes defined in assembly documentation, and the quality processes to ensure compliance with the end applications.

Evaluation of CMs starts with their quality system, which should be at least approved for ISO-9000. Consider how long the CM has held the certification. A supplier that has been certified for more than three years has a demonstrated quality system in place.

Also look at the CM's document-control process to be sure the process is robust and capable of ensuring that the correct documentation is used for each build. Examine the CM's benchmarking data for process controls, look at customer field return data, and determine whether the CM has a system in place for recording this information and using it for continuous process improvement.

Another element that may apply to PCB fabricators is whether the CM has a disaster control plan in place. Should a natural disaster hit the CM, production could be lost for a significant period. A second or backup facility with the same skill set offsets this issue. Another factor to consider is how the CM handles unanticipated problems or delays. Knowing the CM team's experience level, especially between themselves and among like projects, will go a long way in being able to predict how effective and efficient the CM will be at solving problems.

One critical element that must be addressed in an outsourcing strategy is process integrity. Process integrity refers to documented procedures that clearly define the methods to be used, the performance expected, the means to measure that performance, and a metrics system to communicate performance. Most OEMs require every subcontractor of fabricated parts (such as PCB assemblies, cable harnesses, precision machined components, molded plastic parts, etc.) to maintain well-documented manufacturing processes and control mechanisms for monitoring those processes. Visit the CM, review the process documentation, and observe the CM's ability to produce the assemblies before selection.

Process integrity applies to other elements of the operation as well, including order entry, material acquisition and acceptance, and supply-chain management. Although these elements exist in all types of outsourcing, they can be particularly problematic for an EMS provider. As an integrator of other suppliers' components, an EMS provider's success is completely dependent on the success of the supply chain. As such, administrative systems are just as critical as product and process systems. Product-related factors (supply chain, process control, etc.) should be documented in the form of detailed manufacturing instructions, quality control plans, and test protocols that are validated during the first article and preproduction stages of the outsourcing project. Materials turns within the CM's facility are also an indicator of the CM's procurement skills. The higher the number, the better the supply chain control.

The EMS industry is typically divided into two primary segments. In the first segment, low-mix, high-volume CMs are geared up to produce large quantities of the same unit. Typically, they don't have the infrastructure to handle varied and complex products, and much of the production takes place in low-cost regions such as China. These manufacturers are set up to run large volumes of products (e.g. cell phones and consumer electronics) at the lowest cost. High-mix, low-volume CMs tend to make a wide variety of highly complex products (e.g. medical devices, industrial automation, and analytical instruments) in small batches. The very nature of this model requires considerably more-sophisticated infrastructure to support device industry requirements, especially as they relate to regulatory and quality issues. Select CMs that fit the product mix.

Technologically, the CM should have the ability to build the exact type of product needed. Do the PCB assemblies contain microBGAs? Then the supplier should have the appropriate equipment to properly place this type of component and should have X-ray capability for inspection. Are LF components in your design? Then the reflow oven should be LF-compatible, capable of running at higher temperatures (40–50 C) to accommodate LF components. At least a 10-zone reflow oven for any type of LF assembly is recommended. Systems with fewer zones must ramp up too rapidly and can burn off flux and additives before reflow, causing poor solder joint integrity. A list of capabilities that delineates the criteria for a viable CM is the best way to determine the best supplier. The list includes:

- Manufacturing processes
- Equipment available
- Quality certifications/facility approval
- Quality systems
- Process change notice (PCN) process
- Testing capabilities
- Tooling capabilities
- Design support
- Prototype availability
- Volume level
- Conformal coating

- Lead-free
- New product introduction process
- Design validation testing

Having a cultural fit with the CM is also important. Some tangible criteria that may help determine cultural fit include:

- MRP/ERP system capabilities
- Purchasing strength, supplier relationships (onshore and offshore)
- Responsiveness, flexibility
- Experience
- Business practices

6.6.2.2 Auditing a Contract Manufacturer

Introduction This section identifies the activities that are typically explored in a contract manufacturing audit, why each operation is essential, and what some of the commonly occurring defects are at each specific operation. Key steps in a typical SMT manufacturing flow are identified, and why they are audited is explained.

Procedures Specific operational procedures must be available at all steps of the manufacturing process. SOPs, along with weekly maintenance logs identifying what procedures are performed on all systems to ensure the highest performance, must also be available. Finally, operator training data and all quality management systems procedures, such as ISO-9001, should also be available for review.

Incoming Handling of Received Parts and Circuit Boards At the incoming stage, the audit verifies the way that bare PCBs are received from the supplier. Depending on the surface finish, boards may be received in sealed or even moisture barrier bags with desiccant to protect them from oxidation and moisture. Component handling also addresses the MSL of the components to ensure proper storage and handling on the manufacturing floor. MSL parts must be baked according to the requirements when exposed for longer than allowable times. If parts are received in sealed packages, baking does not need to be performed unless parts are left open for longer than the specified time. In addition to controls on the manufacturing line, ESD measures must be implemented at these incoming operations. To help control ESD, the facility should be maintained at a 40–60% humidity level. (Humidity levels outside that range can be tolerated if additional measures are taken.) If standard paper is used at workstations, it should be placed in ESD bags so it can't generate any static. All operators should wear two heel straps and one wrist strap when seated at a workstation. Ideally, all operators wear ESD smocks over their clothing as well. Heel strap testers are recommended to ensure that the heel straps are functional before going onto the manufacturing floor. If possible, the floor in the entire manufacturing area should be a special conductive floor to further minimize ESD.

The storage of solder paste, adhesives, and other process materials must also be audited. This includes evaluating the refrigeration method, storage temperature, length of time the material is out of the refrigerator before it can be used, expiration date, and method for refrigeration after use.

Screen Printing Screen printing accurately and consistently provides the correct volume of solder paste onto each required location (pads and through-holes) and is a critical operation. The audit process inspects the method for dispensing the paste onto the stencil, how the paste is replenished, how long the paste is left on the stencil, and how the paste is disposed of. The method for inspecting print quality is also critical. In some cases, an inspection can be accomplished using the onboard screen-printer technology; but frequently, it is performed using a separate piece of equipment. If a misprint occurs, the PCB must be cleaned in some way to remove the paste. Cleaning typically begins using a manual wipe technique; then the PCB is run through a washer where the apertures are cleaned in the stencil to ensure the next print is good. Auditing this operation is vital. Also review the characteristics of the stencil apertures to identify if enough solder paste is being placed on the pads. The use of appropriate stencil apertures ensures optimum solder joints without causing bridging or opens. Aperture recommendations are provided by component suppliers, equipment manufacturers, and industry standards. They may also be developed in house.

Pick and Place Pick-and-place equipment can take many forms, from tabletop units to multiple in-line systems. The audit must address the procedures for feeder loading and verification and then first-article inspection to ensure that all placements are as programmed. Placement accuracy for the product requirements must also be verified. For area array parts, the vision system operation should be verified to ensure that all balls on BGAs, for example, are present before placement. Also review the methods used for line balancing, as that can impact the overall process flow and exposure time of the solder paste and moisture-sensitive components.

First-Article Inspection/Visual Inspection The audit examines how the first-article visual inspection is performed (e.g. microscope, standard visual with no magnification, 3X ring light, AOI, etc.), as each approach has its limitations. Also, determine who performs the inspection: a line operator, QA inspector, line lead, or variations of all three.

Solder Reflow This operation is also critical. The audit addresses the number of zones in the reflow oven, as this impacts the process window and whether the system can successfully process LF assemblies. In looking for potential problems, the reflow profile is audited to ensure compliance with the solder paste formulation, any special component requirements, and industry standards. The profile should have pre-heat zones, ramp and soak zones, reflow zones, and a cool-down zone for optimum operation. Also audit the degrees C per second change in the ramp-up and -down rates to reduce the possibility of flux exhaustion before entering the reflow zones and thermal shock on exiting the oven. Recommended reflow profiles are provided by component manufacturers, solder paste suppliers, and industry standards such as IPC J-STD-001. Also audit how the various profiles are stored and specifically identified. Improper identification and control can result in the selection of the wrong profile for a specific circuit board. Maintenance and profile-verification techniques should be reviewed to determine whether temperatures are maintained appropriately.

Cleaning Systems Audit the temperature of the cleaning system for optimum cleaning. For aqueous cleaning processes, audit the way deionized (DI) water is brought to the system and recycled (e.g. reverse osmosis system). Observe the resistivity of the system to see if the CM is controlling the cleanliness of the wash system properly. If cleaning agents are used, the concentrations and process controls for them are verified.

Nozzles or jets should be evaluated for impingement angles and pressures. PCB spacing and speeds are reviewed for conveyorized systems. Finally, any material handling equipment used in the process, such as baskets or fixtures, should be evaluated. Cleaning process recommendations may come from the cleaning agent supplier, the cleaning equipment manufacturer, industry guidelines, or developed in-house. Validate processes used to make sure that the correct cleaning parameters are used and controlled.

X-ray Inspection   Most CMs have an X-ray system. Audit the extent to which this tool is used to identify and monitor manufacturing defects. For example, the CM must recognize the limitations of X-ray systems. X-ray equipment can identify shorts under components and gross defects such as missing balls under BGAs but has limited capabilities for identifying horizontal cracks in solder joints.

Automated Optical Inspection (AOI)   Most CMs perform AOI after SMT processes. AOI systems can be programmed to verify presence and absence, polarities and orientations, and verify the markings on most components. They are not typically capable of performing a good evaluation of the quality of all solder joints on the board. So, this operation should be followed by visual inspection for solder joint quality. The audit determines the extent of this operation, the false failure rate, and whether it is being utilized properly.

Through-Hole Component Insertion   Through-hole components are frequently hand-inserted and soldered, although some facilities have automated placement equipment. Through-hole parts are soldered using different solder and flux formulations than the reflow process. The audit must verify that these materials are compatible with the overall process, particularly if an aqueous wash is used. It is also necessary to audit the solder tip selection and solder iron temperature if hand soldering is performed. The following general hand-soldering tips are recommended:

- Use a soldering iron with great thermal recovery. The lower the soldering temperature and the larger the tip, the less heat loss.
- Use a high-power soldering iron.

- Use the largest tip size and shape commensurate with the joint being soldered.
- Use tip temperatures appropriate for the materials used (for example: LF solder: 700 F with 2 to 5 seconds contact time).
- Use of excessive temperatures can damage boards and components.
- Use a portable preheater when more extensive hand soldering is performed.

Secondary Operations The audit should also address all secondary operations. This includes operations like press-fit connector installation, mechanical assembly, and conformal coating. For press-fit connector operations, the fixturing should be audited to ensure that the deflections caused by the operation are minimal. Adequate board support and force control should be present. The process of singulation, also called depanelization, should be audited to ensure that the process does not cause undue deflection of the PCB during the process. Singulation can be accomplished using "mouse bites," a pizza cutter arrangement on V-grooves, or simple breakaways or by being routed. Mechanical assembly can encompass the addition of screws, brackets, heatsinks, spring loads, or other types of mechanical support. Overstresses can be applied through excessive torque, the loads placed by springs, and the weight and placement of heatsinks. All need to be audited to ensure a robust mechanical product. Strain gage assessment is recommended. Excess deflection can result in fractured solder points, damaged laminates, and pad cratering. A conformal coating operation, whether via spray, dip, or brush, should be audited to ensure full and proper coverage. This can be accomplished for some potting materials by inspecting under a UV light.

Rework Any rework operations also need to be audited to ensure that the temperature profiles are not detrimental to the circuit board assembly. Many CMs use higher soldering temperatures to perform this operation rather than mimic the reflow profile. These higher temperatures can damage the circuit board if the decomposition temperature (Td) of the laminate is exceeded. Potential damage includes delamination and copper dissolution. (Copper dissolution is the loss or reduction of copper thickness due to extended exposure to elevated temperatures.) Sensitive components adjacent to the reworked areas may also be damaged. As such, auditing the profile and tip contact time is essential.

Test Operations Manufacturing tests can take on many forms, including functional, flying probe, and ICT. Audit these functions to determine the percent coverage in each test concerning the circuit. Also review the pogo pin style and configuration, since improperly selected and placed pins may cause enough deflection in the PCB to damage laminates and solder joints. Knowing this issue at the design stage is measurably better than determining there is a problem during an audit.

Shipping Proper shipping materials and methods must be verified to ensure that failures are not induced after testing has been completed. Shipping documentation and labeling are also reviewed to ensure that they meet customer requirements.

6.6.2.3 Summary

Designing for the life cycle is truly about designing with the end in mind. DfS incorporates designing for reliability, manufacturability, testability, and serviceability to manage the total cost of ownership and to ensure customer satisfaction. Appropriate life-cycle planning is a core element of a successful Design for Excellence program.

References

Barker, D.B. and Dasgupta, A. (1993). Thermal stress issues in plated-through-hole reliability. In: *Thermal Stress and Strain in Microelectronics Packaging*. Boston: Springer, 648–683.

Caswell, G. (2010). Counterfeit detection strategies: When to do it/how to do it. International Symposium on Microelectronics. doi: 10.4071/isom-2010-TP2-Paper5.

Caswell, G. (2016). Auditing a printed circuit board fabrication facility. DfR Solutions. https://www.dfrsolutions.com/auditing-a-printed-circuit-board-fabrication-facility.

Jones, M.L. (2016). DNA marking technology improves quality through fraud prevention. Defense Logistics Agency (DLA). https://www.dla.mil/AboutDLA/News/NewsArticleView/Article/958928/dna-marking-technology-improves-quality-through-fraud-prevention.

Texas Instruments (TI). (2006). Plastic package moisture-induced cracking. Application report. www.ti.com/lit/an/snoa300a/snoa300a.pdf.

Tomczykowski, W. (2003). A study on component obsolescence mitigation strategies and their impact on R&M. Annual Reliability and Maintainability Symposium. IEEE. doi: 10.1109/RAMS.2003.1182011.

Tulkoff, C. and Caswell, G. (2012). Obsolescence management and the impact on reliability. International Symposium on Microelectronics. doi: 10.4071/isom-2012-WA23.

Tulkoff, C. and Hillman, C. (2012). Procurement-best practices for improving the PCB supply chain. *Printed Circuit Design and Fab-Circuits Assembly* 29 (9): 28.

Tulkoff, C. and Hillman, C. (2013). Best practices for improving the PCB supply chain II: Performing the process audit. SMTAI.

Tulkoff, C., Hillman, C., and DfR Solutions (2013). Reliable plated through via design and fabrication. SMTAI.

7

Root Cause Problem-Solving, Failure Analysis, and Continual Improvement Techniques

7.1 Introduction

Root cause analysis (RCA) encompasses a category of problem-solving methods that focus on identifying the ultimate underlying reason of why an event occurred. Root cause analysis is a generic term for diligent structured problem-solving. Over the years, various RCA techniques and management methods have been developed. All RCA activities are problem-solving methods that focus on identifying the underlying reason of why an undesired issue occurred. RCA is based on the belief that problems are most effectively solved by correcting or eliminating the root causes rather than merely addressing the obvious symptoms. The root cause is the trigger point in a causal chain of events. This trigger point may be natural or manmade, active or passive, initiating or permitting, obvious or hidden. Efforts to prevent or mitigate the trigger event are expected to prevent the outcome or at least reduce the potential for problem recurrence.

RCA is a comprehensive analysis method that identifies the chain of physical and human-related root cause(s) behind an undesirable event. This differs from basic troubleshooting and problem-solving processes that typically seek solutions to specific, relatively simple difficulties. The undesired event may be a product durability failure, a safety incident, a customer complaint, a quality defect, or a result of human error. RCA focuses the corrective and preventive action (CAPA) efforts on the actions

Design for Excellence in Electronics Manufacturing, First Edition.
Cheryl Tulkoff and Greg Caswell.
© 2021 John Wiley & Sons Ltd. Published 2021 by John Wiley & Sons Ltd.

that have the greatest potential impact. It is essential for guiding change management efforts in the most effective and cost-efficient direction.

The goal of RCA is preventing failures before they happen. Prevention can be accomplished via a thorough understanding and analysis of designs, materials, suppliers, and customers. When failures do occur, however, organizations also need methods that enable them to quickly learn from and respond to these events. Effective use of RCA enables the most economical use of resources to achieve these goals. RCA also drives the pursuit of actions most likely to result in success. All companies have constraints on people, money, and time. If an organization doesn't analyze, learn from, and prevent problems, it simply repeats them.

CI is essential in remaining competitive and in becoming and staying world-class. A company that consistently, reliably, and profitably provides the greatest value to customers succeeds. In today's competitive environment, only the survival of the fittest and best is guaranteed. RCA is a critical factor in success.

7.1.1   Continual Improvement

Continual improvement (CI) is the ongoing effort to improve products, services, or processes to advance the goals of an organization, business, or society. It is a never-ending effort to discover and eliminate inefficient process roadblocks and bottlenecks, non-value-added activities, and problems. CI can be either an incremental improvement over time or a breakthrough improvement all at once. It is also referred to as a *kaizen* or "change for the better" process. CI helps maintain a constant and consistent focus on improving products, services, or processes with the overall objective of improving efficiencies, customer satisfaction, and economic performance. Ultimately, CI is viewed as a process for business process optimization and improvement, which is why it is often a priority with organizational leadership. In engineering design and product integrity, CI typically focuses on quality, reliability durability and performance issues. It is based on the concept of obtaining and analyzing feedback from processes, customers, and end-users, and then evaluating performance against organizational goals. Improvements can occur as either gradual small incremental improvements over time or large breakthroughs in a single event.

CI concepts are strongly based on the work of Dr. W. Edward Deming and the Japanese philosophy of kaizen or "improvement," which seeks to identify waste and inefficiencies as well as performance and quality, reliability, and durability (QRD) issues. A key part of the kaizen philosophy is developing a culture where every worker and team member is empowered with responsibility and authority for finding problems and inefficiencies and then developing and implementing improvement strategies. The kaizen CI philosophy differs vastly from the Western, centralized, authoritarian top-down, command-and-control management philosophies that dominated from the Industrial Revolution to the mid-twentieth century. Examples of commonly used CI methods include statistical process control (SPC), Six Sigma Quality, best practices, lessons learned, process optimization, and problem-solving.

7.1.2 Problem-Solving

Problem-solving is an integral part of cognitive thinking and decision-making and is an essential part of daily life. It involves using tools to obtain relevant data, information, and knowledge, creating mental models of situations and how the world works. It further involves making logical connections that lead to the formation of potential solutions and evaluating the potential solutions against goals, constraints, and desires. Problem-solving methodologies include trial-and-error, brainstorming, and RCA. Problem-solving in electronics can be described as high-tech detective work. It has been the basis of engineering since humans first made tools and structures. Lessons learned during the construction of the early step and bent pyramids enabled the ancient Egyptians to later build bigger and better pyramids. Figure 7.1 illustrates the distinctions between standard problem-solving vs. root cause problem-solving.

7.1.3 Identifying Problems and Improvement Opportunities

CI activities begin with identifying problems, non-conformance items, customer or end-user dis-satisfiers, or efficiency improvement opportunities. Sometimes these items can be handled locally or within a department; other times, when greater resources or interdepartmental cooperation is needed, they may need to be escalated to a higher or central authority for prioritization for a corrective action investigation by specialized

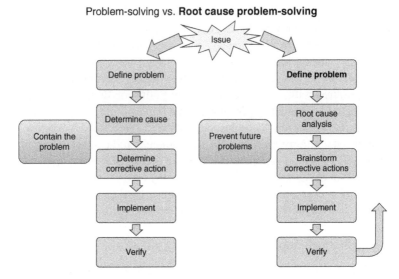

Figure 7.1 Problem-solving vs. root cause problem-solving.

teams of experts. This means an organization must invest in and establish policies and processes for identifying issues, documenting and reporting the issues, and analyzing and prioritizing them for investigation and corrective action. Even within the kaizen approach, where everyone is responsible for CI, there is still a need to document and report problems and CI activities for overall tracking purposes.

The type of issue-reporting methods used depends on the needs of the organization. Issue data may come from warranty data, yield or fall-out reports from statistical process controls, Six Sigma or in-process tests, customer or end-user complaints, field concern reports via company call-in or websites, user-support functions, etc. The most sophisticated companies even monitor or establish a social media website to get product feedback for CI activities.

RCA is not a single methodology; there are many different tools and processes for solving different types of problems. Collectively, they are often referred to as the *quality-reliability toolbox*. Various combinations of problem-solving tools have been developed for different categories of problems:

Examples:

- Manufacturing-related RCA is based in industrial quality control methods.

- Failure RCA methods come from material science failure analysis methods.
- Software failure analysis (FA) is a subcategory involving computer science and programming.
- Forensic engineering requires the use of protocols that are recognized by the legal profession and the court as valid for establishing facts for criminal or civil legal actions.
- Safety-related RCA activities evolved from accident analysis, occupational safety, and health.
- Business process RCA involves accounting and business management methods.

RCA typically involves detective-like investigative skills for collecting facts about how a problem or failure occurred, developing a theory about the root cause that led to the problem, testing the theory to verify that this is the true root cause, and then developing a corrective action plan to resolve the problem or stop the failure from occurring. The following sections will discuss industry standards, process tools, and steps in the RCA as it applies to product and systems engineering.

CI is essential to being competitive and advancing company objectives, and problem-solving is an important tool for CI. RCA is a specific type of problem-solving approach that works to identify not only what undesired event occurred, but also how and why it happened, to prevent a recurrence. FA is a broad subcategory of RCA techniques that can be used when failed or malfunctioning devices are available for examination. FA has many specialties unique to the type of technologies and materials that fail.

7.1.4 Overview of Industry Standard Organizations

Numerous industry standards organizations provide different types of guidance to help facilitate problem-solving and FA. CI requirements are often included in reliability and quality system test books and standards including section 7.2.5.1, "Quality Improvement Program of MIL-HDBK-338B – Electronic Reliability Design Handbook," ISO-9000, and QS-9000.

ANSI
The American National Standards Institute (ANSI) has created standards to ensure the safety and health of consumers and the protection of the

environment. The organization has been in existence for 100 years and has produced over 11,500 standards. ANSI primarily links together organizations like the ASTM, IPC, IEEE, and SAE to generate consensus on standards. SAE ARP 926C-2018, "Fault/Failure Analysis Procedure," is one example of a document that guides performing failure/fault analyses in relatively low-complexity systems.

IPC

IPC-TM-650 test methods include the IPC-TM-650 manual; IPC-9241, "Guidelines for Microsection Preparation"; IPC-9631, "Users Guide for IPC-TM-650"; Method 2.6.27, "Thermal Stress, Convection Reflow Assembly Simulation"; and IPC-9691, "User Guide for the IPC-TM-650, Method 2.6.25, Conductive Anodic Filament (CAF) Resistance Test (Electrochemical Migration Testing)" are particularly useful for FA.

The IPC-TM-650 test methods are free to download and cover the following core areas:

- Reporting and measurement analysis methods
- Visual test methods
- Dimensional test methods
- Chemical test methods
- Mechanical test methods
- Electrical test methods
- Environmental test methods

The IPC-TM-650 test methods do not provide pass/fail or good/bad criteria. Those are up to the user to determine. The methods do outline the industry-accepted tests and the methods used to perform these tests.

JEDEC

JEDEC is the leading developer of standards for the solid-state device industry. All JEDEC standards are available online at no charge at www .jedec.org. JEDEC defines FA as both a methodical process of testing, dissecting, and inspecting a semiconductor device that is suspected of malfunctioning, with the goals of locating the failure site and determining the cause of failure; and the actual investigation to determine the failure mechanism of an electrical or visual/mechanical nonconforming component. Some commonly referenced JEDEC/IPC joint standards for FA include:

- JEDEC JESD 671, "Device Quality Problem Analysis and Corrective Action Resolution Methodology"
- JEDEC JESD 38, "Standard for Failure Analysis Report Format"
- JEDEC JEP 134, "Guidelines for Preparing Customer-Supplied Background Information Relating to a Semiconductor-Device Failure Analysis"
- JEDEC JEP 131, "Potential Failure Mode and Effects Analysis"
- JEDEC JEP 143, "Solid-State Reliability Assessment and Qualification Methodologies"

These documents describe specific guidelines for component-level testing and reporting. The topics covered include test setup and equipment, requirements, and report format. JEP 143 provides an overview of some of the most commonly used systems and test methods historically performed by manufacturers to assess and qualify the reliability of solid-state products.

7.2 Root Cause Failure Analysis Methodology

Root cause analysis is a generic term for diligent structured problem-solving. Over the years, various RCA techniques and management methods have been developed. Five of the most popular are:

1. The 5 Whys
2. The 8D (Eight Disciplines)
3. Shainin Red X (Shainin 2019)
4. Six Sigma
5. Physics of failure / reliability physics

7.2.1 Strategies for Selecting an Approach

There is no one-size-fits-all approach for RCA. Create a toolbox with multiple approaches to choose from, since all the methods have both strengths and weaknesses. When selecting a methodology, consider first the type of problem to be solved. Does it relate to a product, a manufacturing process, a business process, or a combination? Does the problem concern something new or existing? Are there errors to correct, or are improvements being sought? The company or organization culture also needs to be considered. Is the culture formal, flexible, supportive? Tailor the analysis approach to both the problem and the people to maximize the opportunity for success. See Table 7.1, where L = Low, M = Medium, H = High.

7.2.2 The 5 Whys Approach

The 5 Whys approach is a simple problem-solving technique developed by Toyota (Liker 2004) to quickly get to the root of a problem. It simply involves looking at any problem and asking: "Why?" and "What caused this problem?" The answer to the first "why" prompts another "why," and the answer to the second "why" must prompt another, and so on. The rule of thumb is that the "why" question must be asked and resolved at least five times to identify the true underlying root cause of the problem. Toyota's philosophy is that a rush to action addresses only symptoms of a problem and produces only temporary relief. Only after the "true" root cause has been identified can an effective strategy be developed to resolve the issue permanently.

The benefits of this approach are that it is easy to remember, it is simple to apply, and it gets deeper into root cause than many other problem-solving techniques. The 5 Whys is informal and flexible, and it has an open structure with little bureaucracy. Companies can easily adapt it to their own unique needs and cultures. Some disadvantages are that it requires more time to investigate than typical quick-fix approaches. Resolving issues with more than one underlying cause can be difficult to manage. The most challenging part of 5 Whys is asking the right "why" questions. Mistakes in developing and answering a "why" question can quickly derail an investigation. The technique requires both subject matter expertise (SME) and knowledge of cause and effect, which may be lacking in keys areas across an organization. Participants need to know how to follow up on questions to reach useful conclusions. Novices can easily follow an incorrect path. While 5 Whys is an informal and flexible process, little bureaucracy also equates to little guidance.

Table 7.1 RCA method effort comparison.

RCA method	Formality	Complexity	Product	Mfg Process	Business Process	Expertise Level	Cost	Timeline	Flexibility
5 Whys	L	L	H	H	M	L	L	L	H
8D	M	M	H	H	H	M	L	M	H
Red X	H	H	H	H	L	H	H	H	L
Six Sigma	H	H	H	H	H	H	H	H	L
PoF	L	M	H	H	L	H	M-H	M-H	H

Repeatedly asking "Why?" can feel intimidating to the people being questioned. They may fear being blamed for the failure and refuse to cooperate or even fail to disclose helpful information.

Two separate "why" paths should be considered: not only "Why did the problem happen?" but also "Why did the problem escape?" Here is an example:

- What is wrong with what?
- What happened?
- What product(s)?
- Who is involved?
- Who is the customer?
- Who found it – who initially analyzed it?
- Where was/is the problem occurring?
- Where was it made?
- When did it happen? What time? What shift?
- Why wasn't it caught sooner?

7.2.3 The Eight Disciplines (8D)

8D is a problem-solving methodology that emphasizes teamwork. It originated in 1974 as part of MIL-STD-1520, "Corrective Action and Disposition System for Nonconforming Material." Ford introduced and popularized the process within the auto industry in 1987. Initially known as team-oriented problem-solving (TOPS), it evolved into today's widely used 8D process. The underlying philosophy is that when a problem cannot be solved quickly by an individual, a team approach should be considered. Teams are more effective than the sum efforts of individuals working separately. However, it is essential to assign the right members to a team and to support them. Team members need to have the motivation and skills required for each problem, and they must be provided with the time and resources necessary to solve the problem.

The 8Ds drive CI by driving learning across the organization beyond the discrete issue being addressed. The eight disciplines are:

1. Form the team.
2. Describe the problem, and analyze data.
3. Define containment actions.
4. Perform root cause analysis.

5. Choose and verify corrective action.
6. Implement corrective action.
7. Apply lessons learned.
8. Celebrate success, and close the issue.

Figure 7.2 shows the 8D process.

Formal 8D analysis should be considered when any significant deviations occur. Recalls, field failures, and software patches to fix bugs are situations that should require thorough problem-solving. Any time product design, development, or production is interrupted or a corrective action request is issued to a supplier, it should be considered a problem requiring a thorough RCA and permanent corrective action. Process problems and CI projects also benefit from 8D.

To be successful with this technique, organizational support is mandatory. A facilitator or coordinator must assign 8Ds as problems arise. There must be mechanisms for tracking progress and checking for thoroughness in each step. Containment actions and root causes must be implemented and thoroughly verified. The final report should be approved and archived. Finally, the facilitator must summarize and report regularly to management so that the broader lessons learned are assigned and shared

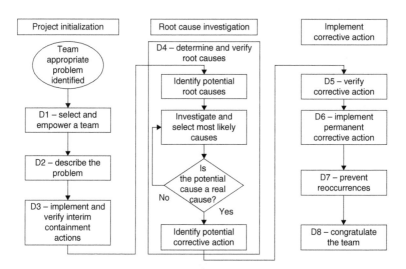

Figure 7.2 The eight disciplines process.

across the organization. 8D problem-solving should never be an individual activity. The team should be cross-functional, with team members from the problem area and other affected areas, both downstream and upstream of the problem. Team members need to have the right technical skills. The most effective team typically has three to six people involved. If the team becomes larger, consider dividing the problem into sub-areas with smaller teams.

Describing the problem accurately requires more time and attention than it usually receives. A well-developed description focuses attention on the right issues and improves the odds of achieving a successful solution. The description leaves a record for others and helps separate the problems from the symptoms. The initial problem description probably will not be complete because the problem is not yet completely understood. Several iterations are usually required as more is learned and more data becomes available. 5 Whys is an excellent tool to use to help craft the problem statement.

An effective problem description details who, what, where, when, why, how, and how many. It contains things like:

- Defect description
- Part numbers or names
- Customer or supplier names
- How the problem was found
- Digital images
- Quantities
- Serial numbers
- History and return material authorizations (RMAs)
- Dates: when found, timeline, build dates
- Change points: engineering changes, equipment, process, supplier, etc.
- Related products or systems
- Data analysis: charts, statistics
- Is/Is not table

Data analysis is needed to describe the problem more fully and eliminate extraneous causes. Data can be either attribute data or variable data but should just represent the facts as known. The is/is not tool is useful once data has been collected. It helps to determine where the problem could be and where it doesn't appear to be. It takes a big problem and helps establish clear boundaries.

Containment actions are brainstormed, selected, and performed in parallel with problem-description activities. The goal is to ensure that no further defects escape. Holds, inspections, rework, and field actions may all be used. Then the focus moves to fixing the immediate issue. This may involve temporary process changes, inspections, and software change and data collection for further analysis.

Once containment has begun, the team continues the RCA. The team must separate symptoms from causes by drilling down from the immediate issue to the root of the problem. They combine experience with the discipline to consider a range of possible causes. The team draws from a suite of tools to perform analysis and closely scrutinize any change points in the process or product. The goal in this stage is to identify the problem, not to jump to a solution. The team is encouraged to consider multiple causes since there is rarely one "smoking gun" that led to the issue. Most problems result from an unexpected confluence of several events. Avoid defining operator error or documentation as root causes. There will always be human errors; focus on how to prevent them. There is also an over-reliance on documentation. Focus instead on how the process can be made obvious and intuitive.

Corrective actions truly prevent the problem from recurring. Corrective actions eliminate the causes, while containment actions eliminate the immediate problem. Corrective actions are permanent and may be implemented immediately or may take substantial time to accomplish. There are three important steps in the corrective action stage: identifying, prioritizing, and verifying. Brainstorm potential corrective actions, and don't rule any out early in the process. Prioritize correction actions by considering effectiveness, cost, and time. Finally, verify all corrective actions and results and look for the introduction of new problems.

Create a robust plan for implementing corrective actions. The action should be broken down into actionable steps with an emphasis on who, what, and when. The plan should remain open until all actions have been completed and verified. Finally, a follow-up verification should be scheduled and performed to make sure the changes were permanent.

Next, the lessons learned must be applied. This step is what distinguishes good problem-solvers from world-class problem-solvers. The concept here is to look at the big picture by addressing systemic issues. The team must consider how the current problem could occur elsewhere in the company in other products, departments, or locations.

This enables an organization to truly achieve real CI. Consider process changes, process development, tool and system changes, and documentation enhancements. Lessons-learned actions are typically beyond the scope and control of the 8D; this is an area where the facilitator or moderator must step in and act. Once this is complete, the team can celebrate their success and close the issue.

7.2.4 Shainin Red X: Diagnostic Journey

Shainin Red X originated in 1948 with Dr. Joseph Juran, Len Seder, and Dorian Shainin. Engineers have long modeled problems using equations using X as a cause and Y as an effect. Juran pioneered the concept of "the vital few and trivial many." Shainin similarly felt that problems were not caused by a little bit of many Xs but, instead, by one dominant X: the Red X. Problems could be solved by finding this Red X.

Shainin showed that cause can exist as an interaction among independent variables. The effect of the Red X is magnified by the square-root-of-the-sum-of-the-squares law and isolates the root cause. He felt his application of statistical methods was more cost-effective and simpler than Taguchi methods. To find the Red X, Shainin swapped pairs of parts between functional and failing parts until the one part responsible for the failure was located. He referred to this as "talking to the parts". Talking to the parts is what differentiates Red X from the Taguchi Design of Experiments. Instead of brainstorming, one or more of four clue-generation techniques are used to determine the cause. There are three key principles in Red X:

- Pareto or 80/20 rule: Few causes to a problem even though there are many possibilities.
- Fact-based decision making: Not opinions, not subjective. Using the actual failures is required.
- Convergent strategies: Effectively cut down the number of possibilities. Avoid the single-good-idea approach.

A Red X statistical "journeyman" or "master" starts the process by organizing a team of problem stakeholders. The team creates a problem definition tree diagram like a fault tree but without the logic symbols. This results in a visual map of the issue or sequence of events that led to failure. It includes relevant issues and realistic contributing factors.

The team uses the diagram as a guide for evaluating the impact of each factor. A progressive search/questioning strategy using a series of yes/no questions about the degree of contribution to failure is used to reduce the field of suspects. Factors that are minor contributors to the outcome are eliminated from serious consideration. The remaining factors in each category line are significant and worthy of detailed statistical evaluations.

Use the remaining significant factors from the problem definition tree to select parts that represent the extremes of the situation. These extremes are called BOBs (best of the best, fault unlikely) and WOWs (worst of the worst, fault most likely). BOB and WOW parts are used to collect problem performance and parameter measurements for statistical relevance evaluations. This identifies causes with the greatest influence on the standard deviation of the outcome. Once the Red X is found, a corrective action strategy can be developed. Then the team can move to the "remediation journey" phase.

The remediation journey phase involves a B (better) vs. C (current) evaluation. After a B corrective action strategy is determined, at least three B units are produced that incorporate the corrective actions. An equal number of C parts are procured. Evaluation tests or measurements are performed on both sets of parts to verify that the identified cause is indeed dominant in the C parts and that a statistical improvement can be measured on the B parts, and to validate the dominant cause and corrective action. Validation is confirmed only if the output values for the B parts and C parts are statistically separate in the expected direction.

The Red X methodology was developed for and is best used for problem-solving on higher-volume manufacturing processes where data is inexpensive to gather, statistics are commonly used, and process changes are complex. It is not very useful for research or product development activities. The benefits of the Red X method are that the progressive search can be more efficient and less subjective than brainstorming. It is effective for variation-reduction problems and is data- and process-driven. Some concerns include the fact that it is a rigid, trademarked process. Red X assumes the existence of only one or two dominant failure causes. The small sample sizes and focus on extremes drive a risk of focusing on process anomalies. Finally, the progressive search requires patience and multiple iterations, so it is potentially a much slower process.

7.2.5 Six Sigma

Six Sigma is a methodology for improving business performance that was developed by Motorola in the mid-1980s. Motorola demonstrated that manufacturing lines with high in-process defects rates requiring rework and repair had higher field failure rates and warranty costs than lines with low repair rates. Low-repair-rate lines also had improved customer satisfaction rates that resulted in better sales. Furthermore, manufacturing lines with better process capability resulted in higher first-pass yield (FPY), making them more efficient and cheaper to operate. This was true even if the better equipment had higher upfront costs since the improved efficiency resulted in higher throughput. The technique enabled quality and reliability professionals to communicate in the native language of executive management: time and money.

Sigma represents a standard deviation in the normal distribution (bell curve). The measurement scale defines how much of a process's normal distribution is contained within the required tolerance limit during the first process pass. Out-of-specification defects are measured in terms of defects per million (DPMM). Processes that operate with a Six Sigma quality capability level produce fewer than 3.4 DPMM for each operation. Data outside the Six Sigma spec limits are "defects." 3.4 DPMM corresponds to 4.5 Sigma. There is an empirical 1.5 sigma shift in the calculation that accounts for real-life process variation. DPMM is related to process operations, not the number of parts produced. The goal of Six Sigma is to create more capable processes that produce a tighter variation spread within the specification limits.

DMAIC (define, measure, analyze, improve, and control) is the Six Sigma improvement system for existing processes and problems along with the processes that fall below specification and yield expectations. DMADV (define, measure, analyze, design, and verify) is the Six Sigma improvement system for developing new processes or products or resolving design-related problems. DMADV is also used in design for Six Sigma (DFSS), a methodology for new product development. Although there are many obvious similarities with the previously discussed 8D and 5 Why problem-solving techniques, there are different definitions and terms and some variations in the statistical tools used.

Six Sigma programs are neither identical nor universal. The basic Six Sigma methodology has many valuable concepts. Be wary of programs where the statistics are too complex to be understood, and programs that endlessly collect data and analyze numbers. Select consultants carefully, since some prefer to work with and train novices who are less likely to question their techniques. Finally, avoid Six Sigma programs where following the process is more important than the outcome.

Benefits of a Six Sigma program include a uniform structure suitable for very large, far-flung organizations. It is also very successful at connecting quality initiatives to business objectives. However, nothing about it is truly different from other statistical process improvement methods. The heavy reliance on complex statistics makes it susceptible to errors and mistakes. It assumes a normal process distribution and is a rigid process. Finally, it can be a profit generator for consultants and certification programs.

7.2.6 Physics of Failure

Physics of Failure (PoF), also known as reliability physics, is a proactive, science-based engineering philosophy of product assurance technology. It is based on a formalized, structured approach to FA that focuses on total learning and not simply on correcting a current problem.

The PoF approach uses science (physics, chemistry, etc.) to capture an understanding of failure mechanisms and evaluate useful life under actual operating conditions. Failure of a physical device or structure can be attributed to gradual or rapid degradation of the materials in the device in response to stress or combination of stresses including thermal, electrical, chemical, moisture, vibration, shock, and mechanical loads. Failures may occur prematurely, such as when a device is weakened by varying processes or by fabrication or assembly defects. They may occur gradually due to wearout issues. Finally, failures may occur erratically when excessive stress exceeds the capabilities or strength of the device.

Knowing how and why things fail is equally as important as understanding how and why things work. Knowledge of how things fail and the root causes of those failures enables engineers to identify and design out potential failure mechanisms in new products. It also helps them solve problems faster. PoF is applicable to the entire product life cycle, including design, development, validation, manufacturing, usage, and service.

PoF techniques were developed to address the limitations of statistics-based reliability predictions. Purely statistical predictions have some fundamental limitations: they should be used only when there is a lack of knowledge of the situation and the knowledge cannot be obtained at a reasonable cost, or when action is necessary and cannot be delayed. The focus of traditional reliability methods on random chance failures conveys a perception that problems and failures are inevitable and unavoidable. When trying to determine a course of action, it is best to acquire knowledge and not blindly use statistics to play the odds.

The PoF approach emphasizes an ordered, understandable, predictable universe of cause-and-effect relationships. The role of RCA problem-solving is to discover, understand, and master the cause and effect relationships. Use RCA to build a compendium of knowledge for future problem prevention as well as for solving today's problems.

In PoF, a failure mode is the effect by which a failure is observed, perceived, or sensed. A failure mechanism is a process (electrical, mechanical, chemical, etc.) that causes failures. *Failure mode* and *failure mechanism* are not interchangeable terms in PoF. The failure site is the location of potential failures, typically the site of a designed-in stress concentrator, design weakness, or material variation or defect. Knowledge of processes and materials is used to identify and prioritize potential failure sites and risks in designs during PoF design reviews.

PoF advantages include that it is knowledge-based, data-driven, and flexible. It is evolutionary in that it builds on previous research and experiments. Disadvantages include the lack of a formal structure, which makes it difficult for novices to use. It is also time-consuming for revolutionary materials and processes since research and characterization are required.

7.3 Failure Reporting, Analysis, and Corrective Action System (FRACAS)

Requirements for a FRACAS system were originally defined in MIL-STD-2155, "Failure Reporting and Corrective Action System (FRACAS)." FRACAS and failure reporting requirements are also included in other reliability and quality system standards including section 8.2, "Failure Reporting, Analysis, and Corrective Action System (FRACAS) and Failure Review Board (FRB)" of MIL-HDBK-338B, "Electronic Reliability Design Handbook."

FRACAS is a closed-loop systematic process for reporting, classifying, and analyzing failures and then planning and implementing corrective actions to resolve them. It was first developed by the US military in the 1980s for identifying and resolving performance, failure, or safety problems with either fielded equipment or systems under development. The basic concepts of FRACAS have since been adapted for use in many different industries.

FRACAS requires processes and policies for:

- Reporting failure issues
- Collecting and evaluating reports by a responsible individual or organization
- Correlating and maintaining databases of related failure issues and equipment
- Prioritizing failures and initiating root cause investigations
- Developing corrective action plans
- Evaluating or testing the effectiveness of corrective action plans
- Implementing corrective action plans into the design or process of the system, or retrofitting fielded equipment
- Presenting reports and proposals to a management authority like a failure review board
- Managing the costs and logistics of the investigation and corrective action phases of the FRACAS
- Documenting the lessons learned from the FRACAS activities for future reuse

Throughout the Design for Reliability (DfR) process and especially during the testing phase, failures of some of the system, hardware, and/or software components are expected. While experiencing a test failure is never pleasant, learning from failure in product development testing is essential for understanding the capabilities and limits of the design, components, and material in the intended application. Finding and fixing problems during testing is much more desirable than dealing with field escapes where issues are far more difficult and expensive to find and fix.

Any failures that occur during the DfR and test process must be well documented to support a detailed root cause failure analysis (RCFA) investigation. Data such as which item failed, how it failed, where it failed, what it was doing when it failed, environmental conditions, etc. should all be collected by the responsible person or group using FRACAS.

Any failures (including test failures) must be fully investigated for root causes, and corrective actions must be effectively implemented to achieve quality, reliability, and durability growth. Furthermore, failures issues need to be converted into lessons learned to avoid repetition of the fault on subsequent products or projects.

An effective failure root cause and corrective action tracking process contain the following key features:

- The failure reporting process requires RCA to be carried out before the failure report can be closed. The conclusions of the analysis should be recorded in the failure record document.
- The failure-tracking process should list the qualifications and experience of the person investigating the failure and as well as of the person reviewing the findings. This could be a graduated experience requirement based on the severity of the failure. If this is done, experience criteria should be stated. The process should attempt to replicate the failure, especially if the exact mechanism is in doubt.
- When appropriate, RCA should identify changes in current procedures to prevent a recurrence. Consider improvements to employee training if a lack of proper design or operating skills was a contributing factor, but recognize that training alone is rarely enough.
- The FRACAS process should be linked to any quality assurance (QA) system used, to ensure that design or operating procedures are changed to support the RCA recommendations and required timelines.

FRACAS should be in place for all projects. Effective use of FRACAS ensures that failures are analyzed properly and that root causes are identified and reported. Recording these reports and lessons learned in a database provides the history to solve issues in current similar projects and prevent issues in future projects.

7.4 Failure Analysis

Implementing a root cause FA process and problem-solving methodology is an integral requirement of a robust design for excellence process. Know the product and process history. Analyze, learn from, and prevent problems, or they will recur.

FA requires a systematic examination of failed devices to determine the root cause of failure. The knowledge gained through FA is used to improve technology, quality, and reliability. FA is primarily associated with the physics and material science of mechanical, structural, and electrical/electronic devices and materials. Software FA is a subcategory involving computer science and programming. Forensics engineering is another subcategory that uses science and technology to investigate materials, structures, products, or components that fail or malfunction to establish facts for criminal or civil legal actions. FA knowledge is vital; an organization that doesn't analyze and learn from problems is doomed to repeat them. FA also helps with prioritization. Most organizations have limited resources and want to spend them in the highest-value areas; FA provides the data needed to make those decisions wisely.

Failure analysis is intended to identify:

- The failure modes (the way the product failed)
- The failure site (where the product failure occurred)
- The failure mechanism (the physical phenomena involved in the failure)
- The root cause (design, defect, or loads that led to failure)
- The recommended failure prevention methods

FA starts with information-gathering to learn about the failure mode being experienced. This is typically accomplished through interviews with key users or members of a product, system, or facility team. The background information about the failure mode is then used to develop a FA plan by choosing the most appropriate FA techniques based on the failure information (failure history, failure mode, failure site, failure mechanism) of the issue of interest.

A physical FA plan always progresses from non-destructive to destructive analysis methods. The objective of the FA plan is to identify the failure signatures of the active failure mechanism. A thorough understanding of the technology, application, and failure physics enables rapid investigation and identification of the active failure mechanism that leads to the root cause.

Before spending time and money on FA, several factors should be taken into consideration. Consider the FA order carefully. Some actions may limit or eliminate the ability to perform further tests. Understand the limitations and output of any tests selected. Use laboratories that can help select and interpret tests for capabilities not available internally. Be careful about requesting specific tests: a better strategy is to describe the problem and define the data and output needed. Pursue multiple courses of action; there is rarely one test or one root cause that will solve a problem. Don't put other activities on hold while waiting for FA results, and understand how long it will take to get results. Most important, consider how the data can be used. How will it help? Can the data be used to drive changes to a course of action, process, or supplier? Don't pursue FA data if it won't help drive action or if there is no control over the path it might take. Some FA is just not worth performing.

Examples of some evaluation techniques used in FA are discussed in the next section. They include:

- Microscopy
  - Optical microscopy
  - Scanning acoustic microscopy (SAM)
  - Scanning electron microscopy (SEM)
  - X-Ray microscopy
  - Infrared imaging
  - SQUID microscopy
- Material and chemical analysis
  - Coefficient of thermal expansion (CTE) measurements
  - Fourier transform infrared spectroscopy (FTIR)
  - Cleanliness and contamination analysis
  - Ion chromatography
  - Micro-hardness measurements
  - Plating composition and quality evaluation
  - X-ray fluorescence
  - Energy dispersive X-ray spectroscopy (EDX)
  - Circuit board solder and soldering quality analysis
    * Tin-lead and lead-free evaluations
    * Solder cross-section and metallographic analysis
    * Evaluation against IPC soldering standards
    * Component pull strength testing
    * Black pad analysis
    * Solder fatigue analysis
    * Dye and pry fracture analysis

- Electrical characterization
  - Impedance measurements (resistance, capacitance, inductance)
  - Hi pot (high potential voltage) limit testing
  - Network analyzer
  - Electrical curve tracers

7.4.1 Failure Analysis Techniques

Understanding the nuances of FA techniques is an integral part of a viable DfM process. However, there is a proper sequence to follow to facilitate usable, actionable results. The best practice approach is to start with a series of non-destructive evaluations (NDEs). This approach is designed to obtain maximum information with minimal risk of damaging or destroying physical evidence. Common non-destructive techniques include:

- Visual inspection
- Electrical characterization
- Time domain reflectometry (TDR)
- Scanning acoustic microscopy (SAM)
- X-ray microscopy
- Thermal imaging (infra-red camera)
- Superconducting quantum interfering device (SQUID) microscopy

Once non-destructive approaches have been exhausted, then using destructive processes is warranted. Destructive techniques include:

- Decapsulation
- Plasma etching
- Cross-sectioning
- Thermal imaging (liquid crystal; SQUID and IR also good after decapsulation)
- Scanning electron microscope; energy dispersive X-ray spectroscopy (SEM/EDX)
- Surface/depth profiling techniques: secondary ion mass spectroscopy (SIMS), Auger
- Optical beam-induced current (OBIC); electron beam induced current (EBIC)
- Focused ion beam (FIB)

- Mechanical testing: wire pull, wire shear, solder ball shear, die shear

Other characterization methods include:

- Ion chromatography (IC)
- Differential scanning calorimetry (DSC)
- Dynamic mechanical analysis / thermo-mechanical analysis (DMA/ TMA)

Let's assess these different approaches in more detail.

7.4.1.1 Visual Inspection

Performing a visual inspection first often provides rapid identification of lot identification (ID) issues. Check for components that might have been part of a prior corrective action and analysis. Look for evidence of counterfeiting. A visual inspection can be done with the naked eye, a magnifying glass, a handheld camera, or a microscope. Digitally document the results. Consider multiple lighting options. Backlighting can provide the ability to look at silhouettes and examine partially transparent samples for internal damage. Ultra-violet (UV) light is useful in detecting contamination and degradation of surface materials such as conformal coating and soldermask. Ball grid array (BGA) perimeter inspections can be done with optical fiber systems capable of inspecting the solder balls on the perimeter of the package. With magnification levels up to 200×, this tool can be useful in identifying common failures in BGAs.

Visual inspection is also good for examining damage such as charring, which is indicative of an electrical short; discoloration of metal, which suggests oxidation, corrosion, or microstructural evolution due to a thermal event; and discoloration of a polymer, which is an indication of contamination or a thermal event. Visual inspection can also identify issues with solder wetting, showing the extent of coverage and the shape of the joints themselves. Component cracking or deformation and surface damage can also be quantified with a visual inspection.

Stereomicroscopy is also useful for detecting tin whiskers since the lighting can be adjusted to influence the degree and orientation of shadowing and make fine whiskers more highly visible.

7.4.1.2 Electrical Characterization

In a FA sequence, electrical characterization may be the most critical step when trying to replicate a failure mode. Determining whether the failure mode is intermittent or persistent is critical, since many intermittent failures are incorrectly diagnosed as "no trouble found" (NTF). Electrical characterization can be accomplished in an ambient environment or in combination with varying environmental exposure: e.g. over a temperature range, over a radiation range, or in a high-humidity environment. None of these added conditions should be severe enough to induce further damage.

Electrical characterization can provide insight by comparing the actual performance of a device against its datasheet specifications. For example, a curve tracer provides plots of voltage vs. current and is valuable in characterizing diode, transistor, and resistor behavior. Time domain reflectometry (TDR) can be used to track a signal along a given path and its return. The measurement of the phase shift of the return signal indicates the potential location of an electrical open.

Electrical characterization may also involve the use of an inductance-capacitance-resistance (LCR) meter, leakage current meters, and four-wire low-resistance meters to measure low milliohms of resistance.

At the bare board level, a bed of nails in-circuit test or a flying probe test may be used to identify defects. These tools are good for identifying shorts and opens and can provide accurate failure-site identification.

Assembled circuit boards can be characterized by a functional test, particularly if the assembly is experiencing a partial failure. Boundary scan (Joint Task Action Group [JTAG]) allows for testing ICs and their interconnections using four input-output (I/O) pins (clock, input data, output data, and state machine mode control). Boundary scan also enables relatively accurate identification of failure sites but is rarely performed on failed units; it is primarily a replacement for in-circuit tests (ICTs).

Oscilloscopes are also a good tool for electrical characterization. They can measure voltage fluctuations as a function of time and are useful in probing operational circuitry.

Overall, using electrical characterization can facilitate the isolation of a defective component without having to remove the component from the assembly. Once isolated, the use of trace isolation can verify the device's performance status.

7.4.1.3 Scanning Acoustic Microscopy

Scanning acoustic microscopy (SAM) is a method for inspecting internal structures through the application of sound waves. The approach requires that the device under test (DUT) be submerged in water and allows for very accurate detection of voids and delaminations. Using different frequencies and transmission modes facilitates the analysis. Using high frequency and a short focal length produces higher-resolution images but has less penetration into the DUT. Lower frequencies with longer focal lengths have greater penetration at a reduction in resolution. Some general rules for parameters are:

- Ultra high frequency (200+ MHz) for flip chips and wafers
- High frequency (50–75 MHz) for thin plastic packages
- Low frequency (15–30 MHz) for thicker plastic packages

The primary SAM transmission modes are pulse echo and through transmission. The pulse-echo mode uses ultrasound reflected from the sample and can determine which interface of the sample is delaminated. This approach requires scanning from both sides to inspect all interfaces and provides images with a high degree of spatial detail. Peak amplitude, time of flight (TOF), and phase inversion measurements can be made.

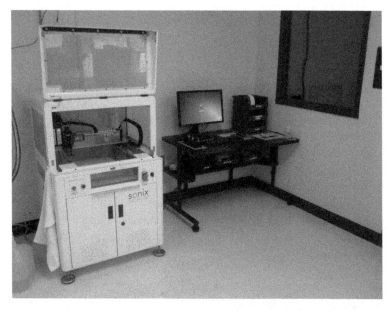

Figure 7.3 Scanning acoustic microscopy system. Source: Ansys-DfR Solutions.

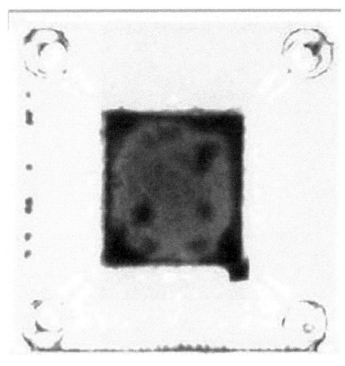

Figure 7.4   Through transmission acoustic microscopy. Source: Ansys-DfR Solutions.

The through transmission mode uses ultrasound transmitted through the sample, where one scan reveals delamination at all interfaces. However, there is no way to determine which interface is delaminated, as this approach has less spatial resolution than the pulse-echo mode.

Figure 7.3 shows an example of a SAM system.

The following images are examples of through transmission (Figure 7.4), peak amplitude (Figure 7.5), and phase inversion modes (Figure 7.6) of acoustic microscopy illustrating the various issues that can be readily identified using this FA tool.

7.4.1.4   X-Ray Microscopy

The industry has continued to make electronic components smaller and smaller. High-resolution X-ray provides the ability to identify cracks, open solder joints, and voids using this inspection and FA technique.

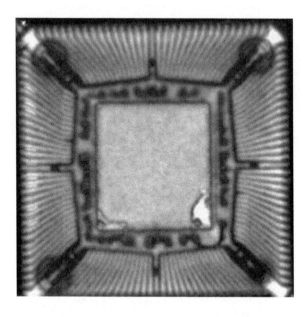

Figure 7.5
Peak amplitude acoustic microscopy. Source: Ansys-DfR Solutions.

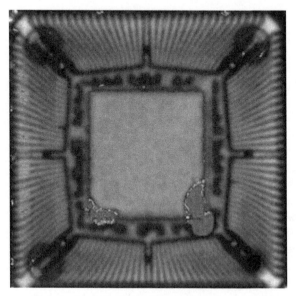

Figure 7.6
Phase inversion acoustic microscopy. Source: Ansys-DfR Solutions.

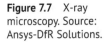
Figure 7.7 X-ray microscopy. Source: Ansys-DfR Solutions.

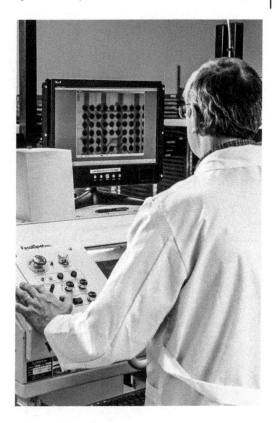

Once the specimen is located under the X-ray beam, it is necessary to adjust the current, acceleration voltage, and gain to achieve the desired image contrast required to view the features of interest. Figure 7.7 illustrates a typical X-ray analysis system.

7.4.1.5 Thermal Imaging
This technique uses thermal imaging in the long infrared range to produce images of that radiation. The amount of radiation created by a packaged device on a printed wiring board increases with temperature, permitting the thermography to assess the variations in temperature in the image. When viewed, hotter objects are more visible against a cooler background. Figure 7.8 is an example of a thermal image demonstrating the variance visible to the user.

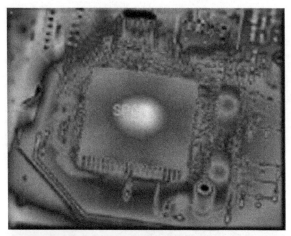

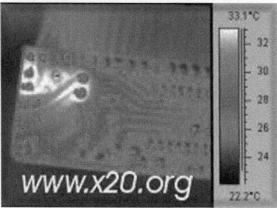

Figure 7.8
Thermal imaging. Source: Ansys-DfR Solutions.

7.4.1.6 SQUID Microscopy

SQUID microscopy is a FA technique that uses weak magnetic fields to assess the location of a short in a circuit board or device. The system can identify extremely low levels of current flow so that shorts can be located and identified without damaging the evidence of the short. Figure 7.9 is an example of a SQUID system.

7.4.1.7 Decapsulation

Once the tools that can perform non-destructive analyses have been exhausted in support of a DfM analysis, it is often necessary to progress

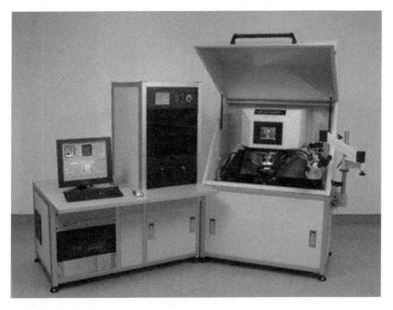

Figure 7.9 Superconducting quantum interfering device microscopy. Source: Ansys-DfR Solutions.

to using destructive FA techniques. One technique is the use of decapsulation equipment, as shown in Figure 7.10. The concept of this analysis is to expose the internal components of a packaged device or circuit board. Using decapsulation techniques to carefully open a component so that detailed inspections can be performed is a vital tool for DfM assessments of component viability and reliability.

During any decapsulation process, you must ensure that no damage occurs to the die or bond wires so that the subsequent visual inspection can ascertain any visible issues en route to determining the root cause of failure.

7.4.1.8 Cross-Sectioning

Another commonly used destructive FA technique is cross-sectioning. This approach exposes a section or plane of a component or circuit board to permit a more detailed assessment of the section. The process usually entails encapsulating the area, grinding and polishing the specimen to the specific plane in question in the device.

Figure 7.10 Decapsulation system. Source: Ansys-DfR Solutions.

Place the specimen into the encapsulant in such a way that damage from the grinding and polishing operations are minimized. Use a vacuum to remove air bubbles from the encapsulant, which would have a negative visual impact on the cross-section. Once the sample is encapsulated for protection, the grinding operation is initiated using different grit silicon carbide paper to slowly grind near the desired location. Several polishing procedures can be used depending on the sample composition and degree of polish needed (i.e. based on what the cross-section needs to show to meet project requirements). The grit of the silicon carbide papers depends on whether the surface to be inspected is solder, silicon, fire retardant (FR)-4 laminate, gallium arsenide, or ceramic. Take care to select the right materials to achieve a highly polished sample for inspection. Figure 7.11 shows a typical polishing operation.

Cross-sections can identify or show both good and bad characteristics of a sample being inspected. Figure 7.12 shows an example of a cross-section showing a solder ball from a BGA device.

7.4.1.9 Scanning Electron Microscope / Energy Dispersive X-ray Spectroscopy (SEM/EDX)

X-rays are produced in an electron microscope whenever the primary electron beam or back-scattered electrons strike metal parts with enough

Figure 7.11 Cross-section polishing. Source: Ansys-DfR Solutions.

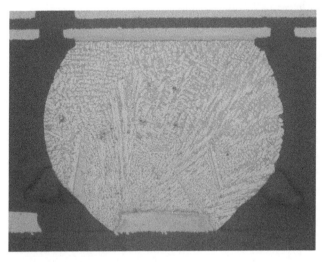

Figure 7.12 Cross-section of a BGA.

energy to excite continuous and/or characteristic X-radiation. In terms of X-ray hazards, two aspects are important: (i) the composition of the parts that are struck and their efficiency as X-ray sources and (ii) the effectiveness/integrity of the shielding provided by the metal casing of the microscope around these.

For the best imaging, the sample needs to be placed as close to the electron source as possible. The charging of non-conductive surfaces is controlled by adjusting the pressure, where a pressure of 60 to 80 mPa is typically enough for most non-conductive samples. If the surface is conductive, then high vacuum mode can be used. Select an aperture to facilitate the magnification needed.

When using EDX mode, it is helpful to understand that the best EDX settings are not the best SEM imaging settings. The best practice is to capture SEM images with optimum SEM settings and then change the settings to optimize for EDX. However, at relatively low magnification, acceptable SEM images can be acquired at optimal EDX settings, which may save analysis time.

The typical working distance for EDX is 10.0 mm. The best practice is to set the working distance to 10.0 mm and then move the stage until the area of interest is in focus. A final focusing may cause the final working distance to be between 9.9 and 10.1 mm.

The EDX detector views the sample from the top of the image when looking at SEM and EDX images (and at approximately 45 degrees up from the plane of the mount). Tall objects create shadows where nothing can be detected, and it may not be possible to detect anything in crevices.

EDX maps that show the presence of a specific element should typically contain at least 5 million counts. In general, EDX maps and spectra should be allowed to collect data long enough such that the peaks of elements of interest are obviously above baseline and are not changing significantly.

If trace elements in maps and spectra are important, their total count may need to be significantly higher than the main elements present. It is then up to the operator to analyze for element identification and compare the fitted spectrum with the collected spectrum. Figure 7.13 illustrates a typical SEM/EDX system and Figure 7.14 shows examples of typical elemental spectra.

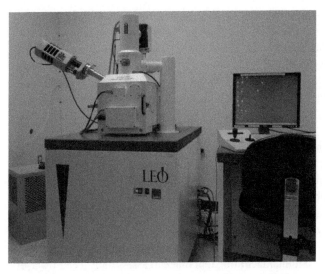

Figure 7.13 Scanning electron microscope system. Source: Ansys-DfR Solutions.

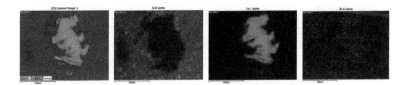

Figure 7.14 SEM EDX spectra. Source: Ansys-DfR Solutions.

7.4.1.10 Surface/Depth Profiling Techniques: Secondary Ion Mass Spectroscopy (SIMS), Auger

SEM is also a viable tool for other methods of analysis to facilitate a DfM assessment. One method is SIMS. This FA technique is used to assess the composition of solid surfaces by sputtering the specimen surface with a focused ion beam and then collecting and analyzing the emitted secondary ions. This approach allows you to determine the molecular composition of the surface down to a depth of 1 to 2 nm. Due to the sensitivity of SIMS, it can provide elemental detection levels down into the parts per billion range.

Similarly, the SEM Auger technique is also a valuable FA tool. This approach uses energetic electrons emitted from an excited atom after internal relaxation events. The analysis of the sample depends on measuring the yield of Auger electrons during a probing activity. The Auger technique examines the radiative and non-radiative decay processes to ascertain the primary de-excitation path, helping primarily in evaluating the lighter elements.

7.4.1.11   Focused Ion Beam (FIB)

FIB is used primarily in the FA of semiconductor die. Using a focused beam of ions, the system can be used for defect analysis, circuit modification, photomask repair, and sample preparation of site-specific areas of integrated circuits. FIB is a destructive technique because when high-energy gallium-arsenide ions impact the sample, they sputter atoms from the surface. Due to this sputtering, the FIB is typically used as a micro and nano machining tool that can assess an IC layer by layer.

7.4.1.12   Mechanical Testing: Wire Pull, Wire Shear, Solder Ball Shear, Die Shear

During a DfM analysis, it is often necessary to assess the wire bonds, die shear, or solder balls on a BGA or other devices. The system used should be capable of applying forces ranging from 1 gram-force to 200 kilograms-force while simultaneously controlling the temperature. Multiple cameras are also needed to obtain the best imaging during the application of stress. Figure 7.15 is an example of such a system.

Other characterization methods are discussed next.

7.4.1.13   Fourier Transform Infra-Red Spectroscopy FTIR

FTIR is a technique to analyze the infrared spectrum of absorption or emission of a solid, liquid, or gas. The raw data is subjected to a mathematical process to be assessed as a spectrum. The system analyzes the amount of light emitted at each wavelength. However, FTIR allows light from many different wavelengths at one time and measures the total beam intensity. Then the beam is modified for a different set of wavelengths to obtain a second data point. Repeating this process many times and then performing computer analysis creates the output information. Figure 7.16 is an image of a typical FTIR system.

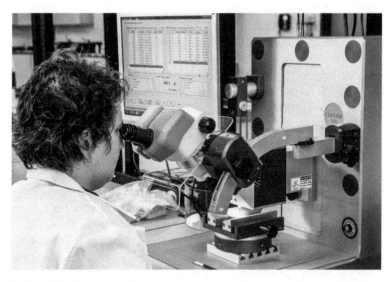

Figure 7.15   Xyztec combination wire bond and shear tester. Source: Ansys-DfR Solutions.

Figure 7.16   Fourier transform system. Source: Ansys-DfR Solutions.

7.4.1.14 Ion Chromatography

The printed circuit board industry has long been interested in the ionic cleanliness of printed board surfaces and its correlation with corrosion, electrochemical migration, dendritic growth, and subsequent opens, leakage current, or shorting during testing and in the field. Initial methods for cleanliness evaluation included resistivity of solvent extract (ROSE) that measured the conductivity of a solution after flowing it over a surface of interest. A major disadvantage of this technique was its inability to detect specific ionic species generating conductivity.

Ion chromatography (IC) has become an important technique for the evaluation of ionic cleanliness. This technique, which detects individual ions, allows quicker troubleshooting of contamination sources and better predictions about the detrimental effects of each ionic species by themselves. IC is a form of high-performance liquid chromatography (HPLC) and works with aqueous sample solutions. It can measure concentrations of major anions, such as chloride and bromide, as well as major cations such as sodium, ammonium, and potassium in the parts-per-billion (ppb) and low parts-per-million (ppm) range. Concentrations of weak organic acids (WOAs) can also be measured through IC.

IC uses ion exchange columns through which ionic species interact with a packed resin. The separation power is based on many factors, including the ionization equilibrium (pKa) and physical size of each ion as well as the resin type and ionic side change density within the resin. A conductivity detector is used for detection as each species leaves the column. Columns and systems are customized and optimized for specific applications and specific ions and groups of ions. Sample solutions pass through a pressurized chromatographic column where ions are absorbed by charged chemical groups attached to the column resin. As an ion-extraction liquid, known as *eluent*, runs through the column, the absorbed ions begin separating from the column. The retention time of different species determines the ionic concentrations in the sample.

The optimum approach uses three separate columns and systems to detect anions, cations, and WOAs. Cations and anions have a minimum detection limit of 50 to 100 ppb (50–100 micrograms/L). Weak organic acids are more difficult to detect because of their lower conductivity values and therefore have a higher detection limit of 100 to 500 ppb (100–500 micrograms/L), depending on the specific acid.

Blanks and National Institute of Standards (NIST)-traceable standard solutions followed by the sample extraction solutions are analyzed using

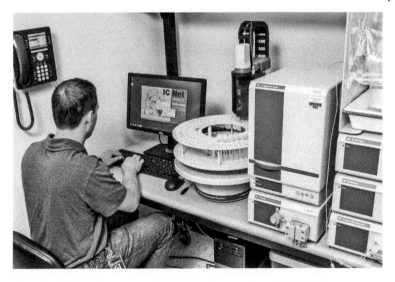

Figure 7.17   Ion chromatography system. Source: Ansys-DfR Solutions.

the IC system. IC values (in ppm) are converted to micrograms/square inch, and board surface areas are calculated using IPC-TM-650, method 2.3.28. Panel surface areas are calculated by length times width times two to account for both sides. Any large holes or open areas are subtracted from the total area. Populated boards are calculated the same way, but with 10% added to account for component surface area. Figure 7.17 shows a typical IC system.

7.4.1.15   Differential Scanning Calorimetry (DSC)
DSC is a FA technique that measures the thermal properties of materials to establish a correlation between the temperature and the physical properties of the material. Essentially, the process involves a thermal analysis that ascertains the temperature associated with material changes as a function of time and temperature. The approach can determine the phase transitions of a material and can assess the melting points and transition temperatures of solutions, solids, or mixed-phase compositions. For basic operation, energy is input to a sample cell and a reference cell, and the temperatures of both cells are raised simultaneously. The difference in the energy required to match the two samples is the amount of excess heat absorbed or released by the sample during the test.

7.4.1.16 Thermomechanical Analysis / Dynamic Mechanical Analysis (DMA/TMA)

TMA and DMA are similar methodologies for characterizing the thermal properties of a material as a function of time, frequency, stress, atmosphere, or multiple stresses from a combination of these parameters.

DMA cyclically applies a small sinusoidal deformation to a test sample of known geometry to apply either stress or strain. The various parameters can then be analyzed, with the change observed being the deformation noted. TMA applies a constant static force to material and observes changes in the material due to changes in temperature and time. The output is dimensional changes. In a FA, the glass transition temperature (Tg) of a circuit board laminate is often determined using this approach.

7.4.1.17 Digital Image Correlation (DIC)

Similarly, DIC can be used to determine strain and displacement of a sample under test. During FA, this tool is often used to evaluate crack propagation and material deformation. The process works by comparing images of a test sample at different stages of deformation and presenting them in either 2D or 3D strain maps.

The process involves covering the test sample with a thin base of flat/matte white paint. Other colors or no paint can be used in some cases, depending on the sample. Once the paint has dried, a contrasting speckle pattern is applied using one of the following techniques: spray can, ink roller, ink stamp, airbrush, brush bristle, or permanent marker. An ideal speckle pattern has 50/50 coverage of white "background" and speckles, with a random pattern and uniform speckle size. Typical speckle sizes should be 5–20 pixels per speckle (fewer than 3 pixels per speckle can cause problems). Next, determine the desired image capture approach: a custom image-capture schedule, fixed time increments between images, or manual capture. Program the thermal chamber to the desired temperature profile and initiate the timed image capture, followed shortly by initiating the test. Figure 7.18 illustrates the typical setup for a DIC test.

7.4.1.18 Other Simple Failure Analysis Tools

Some simple, inexpensive FA tools include the following:

- Portable preheaters allow for controlled heat for component rework and replacement at a debug bench.

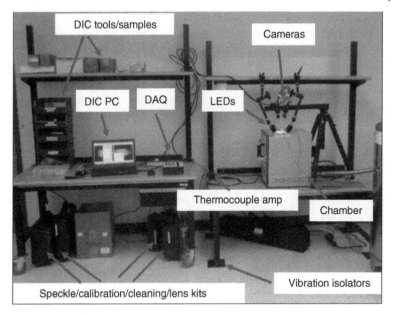

Figure 7.18 Digital image correlation setup. Source: Ansys-DfR Solutions.

- Cold or freeze spray and hot plates induce failures at temperature extremes.
- Temperature and humidity indicator strips monitor environmental conditions.
- Flexible light sources and mirrors make it easier to look under and around areas on a product.
- Dye and pry capability allows quick (destructive) inspection for cracked or fractured solder joints under leadless components like BGAs and QFNs.

7.4.2 Failure Verification

After FA techniques have identified the failure mode, site, and signatures, the failure mechanism(s) that caused the failure must be identified to determine the root cause. Many times, it will not be possible to identify a single specific failure mechanism. However, theories may be developed about potential alternative failure mechanisms. You then need to perform additional evaluations or experiments to identify the specific failure mechanism that is causing the problem. Typically, this involves trial builds with a seeded stock of suspected defects or accelerated stress or life testing that attempts to duplicate the conditions suspected of causing the failure.

The Shainin Red X statistical problem-solving process requires that the statistical or probabilistic root cause projection be confirmed through testing that demonstrates that the failure event can be turned on and off by applying or removing the Red X root cause. In actual practice, this is hard to do, particularly in cases with rare failure situations that affect a very small percentage of production or a small portion of a population of fielded systems. Confirmation can also be difficult in a situation where a complex chain of events or a combination of conditions is needed to trigger a failure mechanism. In these situations, modeling and simulations can be effective in failure verification.

7.4.3 Corrective Action

Identifying a true root cause is essential for developing an effective corrective action plan. The goal is to eliminate or mitigate the root causes responsible for triggering and propagating the failure mechanism that leads to the failure, problem of interest, or non-conformance.

Once the true root cause or root cause candidates have been identified, they must be communicated to the correct person so that effective design, process, or operating changes or other countermeasures can be developed and deployed. The Shewhart or Deming cycle, also known as the plan–do–check–act (PDCA) process, as shown in Figure 7.19, is often used to develop and validate the corrective action strategy. The four steps in the PDCA process are:

1. *Plan*: Identify and analyze the problem to find the root cause.
2. *Do*: Develop a potential solution, and perform small-scale testing to verify the effectiveness.
3. *Check*: Measure how effective the test solution was, and analyze the results to determine if it could be made even better. Identify and adjust to correct for differences between actual and planned results. The check step can also be used to adjust the implementation scheme and determine or verify the cost of the corrective action.
4. *Act*: Formally implement the final solution, and track the actual performance over time to verify that the desired level of effectiveness is

Figure 7.19 Plan-do-check-act process.

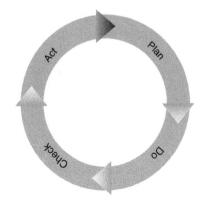

achieved. If desired outcomes are not being fully realized, the RCA investigation may need to be reopened.

7.4.4 Closing the Failure Report

Implementing a corrective action plan does not complete an RCA investigation. The RCA is not completed until the issues, corrective action plan, and outcome are fully documented in a report. The report should be stored in a searchable database so that the project information can be referenced and reused if a similar problem arises in the future or in a different area of the organization.

Some organizations choose to break down findings into individual lessons-learned documents. A *lesson learned* is defined as knowledge or understanding gained by experience. The experience may be positive, as in a successful test or mission, or negative, as in a mishap or failure and how it was dealt with or resolved. The lessons-learned document should identify a specific design, process, or decision that reduces or eliminates the potential for failures or reinforces a positive result. Best-in-class companies are noted for their ability to document and widely communicate lessons-learned information across their organizations.

The 8D problem-solving process includes a living report format that documents the key steps of the root cause investigation as the investigation progresses. The document is used to provide management status reports and to keep team members and their organizations aligned as the RCA progresses. It also provides the formal final documentation report when the RCA project is complete.

7.5 Continuing Education and Improvement Activities

Getting all workers or team members at all levels involved in and able to support CI activities requires training. For an organization that does not already practice CI and RCA, this training must first focus on establishing an organizational culture committed to CI. Cultural change does not occur easily or quickly for people or organizations. A change-agent approach is required, where executive leadership first makes a strong case and communicates why culture change is needed. The organization also must decide on the CI and RCA processes that relate best to what the company does and then start to build an infrastructure that supports those processes.

Once culture change has been initiated, training on the selected CI processes and infrastructure tools is required. Several levels of training are typically needed: an executive overview, a review for the general employee or user population who need to be aware of what is going on, and finally detailed training for the primary participants in CI or RCA activities.

Training for the CI practitioners should follow a logical progression that starts with the overall CI process and then covers the selected approach to FRACAS, data analysis, root cause problem-solving, and problem analysis tools and test methods. Infrastructure resources should be included in the training program for the skill set they support rather than presented at separate training events.

It is also important to include training on the proper way to write FRACAS, RCA, and investigation or test reports. Since these reports might be viewed by government or industry regulators or be subpoenaed for a court case, they need to be factual and accurately define risk issues without resorting to hyperbole, exaggeration, or overstatement.

7.6 Summary: Implementing Root Cause Methodology

RCA is an effective and efficient way to permanently fix and prevent problems. RCA benefits businesses by stopping failures before they happen.

By incorporating a thorough understanding of design and materials and suppliers and customers with a root cause problem-solving methodology, companies learn quickly from failures when they occur and then ensure that they don't occur again. Striving toward world-class problem-solving and FA maximizes the opportunity for success while economizing resources. However, this is a continual journey. Problem-solvers must constantly seek out new tools and techniques. Fortunately, a wide range of options is available for every budget. Conferences, webinars, courses, books, and professional societies like ASQ, SMTA, and IEEE are great resources.

However, internal options are often overlooked. Internal electronic bulletin boards, posters, and demonstrations showing successful root cause projects are great ways to demonstrate the cost savings achieved by RCA. "Brown bags" and "lunch and learns" from internal experts, customers, and suppliers are also effective strategies for staying current with the latest information.

References

Liker, J. (2004). *The Toyota Way*. McGraw-Hill Education.

Shainin. (2019). Preventive problem solving – resilient design and development. https://shainin.com/what-we-do/problem-solving/preventive-problem-solving.

8

Conclusion to Design for Excellence: Bringing It All Together

8.1 Design for Excellence (DfX) in Electronics Manufacturing

Designing for Excellence (DfX) is a methodology where multiple functional groups with knowledge of different parts of the product lifecycle advise the design engineering functions during the design phase. It is also the process of assessing issues beyond the base functionality, where *base functionality* is defined as meeting business and customer expectations of function, cost, and size. Key elements of a DfX program include designing for reliability, manufacturability, sourcing, and the environment. DfX efforts require the integration of product design and process planning into a cohesive, interactive activity known as *concurrent engineering*.

This book's seven chapters have detailed the concept of designing a product with the best reliability so that you, as the user, can provide your customers with the life expectancy that is required.

8.2 Chapter 2: Establishing a Reliability Program

This chapter provided insight into the methodology used to develop a viable reliability program within your organization. A comprehen-

Design for Excellence in Electronics Manufacturing, First Edition.
Cheryl Tulkoff and Greg Caswell.
© 2021 John Wiley & Sons Ltd. Published 2021 by John Wiley & Sons Ltd.

sive, well-thought-out reliability program ensures that companies achieve their quality, reliability, and customer satisfaction targets on time, on schedule, and within budget. Reliability is the measure of a product's ability to perform the specified function at the customer (with their use environment) over the desired lifetime. By definition, reliability is specific to each application – there is no one-size-fits-all definition.

There are also no universal best practices. Every company must choose the appropriate set of practices and implement a program that optimizes return on investment in reliability activities. Reliability is all about cost-benefit; even the most successful organization has a finite budget. Since reliability activities are not a direct revenue generator, they are strongly driven by cost. By increasing efficiency in reliability activities, companies can achieve a lower risk at the same cost. And addressing reliability during the design phase is the most efficient way to increase the cost-benefit ratio.

Best-in-class companies have a strong understanding of critical components. Component engineering typically starts the process through the robust qualification of suppliers and supplier-provided parts. They only allow bill of material (BOM) development using an approved vendor list (AVL). Most small to mid-size (and even large) companies do not have the resources to assess every part and part supplier. Best-in-class companies focus resources on those components critical to the design. Component engineering, in partnership with design engineers, also performs tasks to ensure the success of critical components. This includes the design of experiments, test to failure, modeling, supplier assessments, incoming parts inspection, etc.

Desired lifetime and product performance metrics must be identified and documented. The desired lifetime may be defined as the warranty period or by customer expectations. Some companies set reliability goals based on survivability, which is often bounded by confidence levels such as 95% reliability with 90% confidence over 15 years. The advantages of using survivability are that it helps set bounds on test time and sample size and does not assume a specific failure rate behavior (decreasing, increasing, steady state).

8.3   Chapter 3: Design for Reliability (DfR)

Chapter 3 offered a myriad of topics that are critical to the performance of a DfR assessment of your product. The foundation of a reliable product is a robust design that provides margin, mitigates risk from defects, and satisfies the customer. Ensuring the reliability of electronic designs has become increasingly difficult due to the increasing complexity of electronic circuits, increasing power requirements, the introduction of new component and material technologies, and the introduction of less robust components. The result is multiple potential drivers for failure. Most mistakes in DfR occur due to an insufficient exchange of information between electrical design and mechanical design, weak understanding of a company's limitations, and failure to incorporate customer expectations (reliability, lifetime, use environment) into the new product development (NPD) process.

Design for Reliability is a process for ensuring the reliability of a product or system during the design stage before physical prototypes are made. Crucial elements of a DfR program include setting specifications at the concept and/or block diagram stage, selecting suppliers and parts, and considering wearout mechanisms and physics of failure for the use environment.

Important specifications to capture at the concept stage are reliability expectations, the use environment, and any dimensional constraints. Failure to capture and understand product specifications at this stage lays the groundwork for mistakes at schematic entry and layout. Reliability expectations and goals include the desired product lifetime and product performance needs. *Desired product lifetime* is defined as when the user will be satisfied and should be actively used in the development of part and product qualification plans. *Product performance* can be defined as returns during the warranty period or survivability over the lifetime at a set confidence level. Avoid the use of MTBF (mean time between failures) or MTTF (mean time to failure) as reliability metrics: these calculations tend to assume that failures are random in nature, so they do not motivate failure avoidance. Manipulating these numbers is easy due to the use of multiple quality-adjustment factors. MTBF and MTTF are commonly misinterpreted and are a better fit for logistics and procurement activities.

After reliability expectations have been identified, the next step is to define the use environment. The two most common approaches use industry standards or actual measurements of similar products in similar environments. Using industry standards requires no additional expense since they are free to download. There is agreement on their use throughout the industry; and they are often quite comprehensive. However, most are more than 20 years old and significantly overshoot or undershoot the actual conditions. The second approach is based on actual measurements of similar products in similar environments. For this approach, designers need to determine both the average and realistic worst-case environments. Here, it is critical to identify all failure-inducing loads and all environments, including manufacturing, transportation, storage, and field loads. Transportation and storage are frequently overlooked and may be more severe than some of the use environment conditions, particularly in terms of peak temperature and vibration.

Performing all the design stage work before the physical prototype stage is self-explanatory. Some claim that activities after physical prototyping, such as highly accelerated life testing (HALT), root-cause analysis, and reliability growth, still provide some value. The counter to that argument is that with all the simulation tools available, DfR activities should strive toward optimizing reliability at the design stage. Increasing demand for efficiency is pushing product development from "How many heads per design?" to "How many designs per head?" Within this reality, traditional DfR activities that take up valuable resources, delay the market launch, or provide only vague guidance are increasingly out of favor.

Finally, DfR cannot be an island unto itself. Reliability, manufacturability, testability, functionality: all these disciplines significantly influence each other, and a decline in one discipline has a cascading effect on eventual performance and ultimately failure to meet reliability goals. DfR should be one portion of a larger DfX strategy that is in line with the larger goals of the corporation.

8.4   Chapter 4: Design for the Use Environment: Reliability Testing and Test Plan Development

Product test plans are critical to the success of a new product or technology. Plans need to be stressful enough to identify defects and yet show a correlation to a realistic environment. The recommended approach is to employ both industry standards and reliability physics (RP). This approach results in an optimized test plan that is acceptable to management and customers.

Utilizing RP to facilitate the design, performance, and resulting interpretation of accelerated life tests, starting at the design stage of a product and continuing throughout the life cycle of the product, is the ideal methodology for creating a viable test plan.

It is useful to start with industry specifications, modify them as necessary, and then tailor the test strategy specifically for the product, set of materials, use environment, and reliability metrics required. However, industry testing often falls short due to a limited degree of mechanism-appropriate testing. Often, mechanism-specific coupons are used, which do not reflect the real product, and the test data is hidden from end-users.

Similarly, JEDEC standard tests are often promoted to original equipment manufacturers (OEMs). These tests, typically lasting for 1000 hours, hide wearout behavior. The use of a simple activation energy along with the incorrect assumption that all mechanisms are thermally activated can result in an overestimation of failures in time (FIT) by 100× or more.

Creating a test plan that utilizes RP as its basis is the most cost-effective method for determining a product's ability to perform over its intended lifetime. Again, RP is defined as the use of science (physics, chemistry, etc.) to capture an understanding of failure mechanisms and evaluate useful life under actual operating conditions. Armed with this knowledge, an engineer can evaluate design and performance, and correctly interpret the results of accelerated life tests, starting at the design stage and continuing throughout the lifecycle of the product.

Failure of a physical device or structure (i.e. hardware) can be attributed to the gradual or rapid degradation of the material(s) in the device in response to the stress or combination of stresses the device is exposed to. Common stresses are thermal, electrical, chemical, moisture, vibration, shock, mechanical loads, etc. Failures may occur prematurely, gradually, or erratically.

A viable test plan has defined test objectives that may include test by comparison, qualification testing, validation testing, research, compliance, regulatory testing, and failure analysis. Each of these objectives has a specific goal in mind for the test approach, so having a clear understanding of them is a critical first step.

8.5 Chapter 5: Design for Manufacturability

By following the approach detailed in this chapter, a viable DfM process can be performed early in the design cycle, offering a significant improvement to the product and enhancing its ability to achieve its required life expectancy.

In the electronics industry, the quality and reliability of any product are highly dependent on the capability of the manufacturer, regardless of whether it is a contract manufacturer or a captured shop. Manufacturing issues are one of the top reasons companies fail to meet warranty expectations. These issues can result in severe financial pain and eventual loss of market share. What a surprising number of engineers and managers fail to realize is that focusing on manufacturing processes addresses only part of the issue. Design plays a critical role in the success or failure of manufacturing and assembly.

DfM is the process of ensuring that a design can be consistently manufactured by the designated supply chain with a minimum number of defects. DfM requires an understanding of both best practices and the limitations of the supply chain. The goal is to ensure that the critical objectives of cost, quality, reliability, regulatory compliance, safety, time-to-market, and customer satisfaction are known, balanced, monitored, and achieved.

Building on DfR, successful DfM efforts integrate product design and process planning into a cohesive, interactive activity known as concurrent engineering. If existing processes are to be used, new products must be designed to the parameters and limitations of these processes regardless of whether the product is built internally or externally. If new processes are to be used, then the product and process need to be developed concurrently and carefully, considering the inherent risks associated with anything new.

The foundation of a robust DfM system is a set of design guidelines and tasks to help the product team improve manufacturability, increase quality, reduce lifecycle cost, and enhance long-term reliability. The guidelines must be customized to each company's culture, products, and technologies and based on a solid understanding of the intended production system.

You need to know and understand problems and issues with current and past products concerning manufacturability, delivery, quality, repairability and serviceability, regulatory issues, and recalls. Implement strategies to address and prevent recurrence of mistakes using an effective system for capturing and disseminating this historical knowledge throughout the organization. At an absolute minimum, hold brainstorming sessions after product launch to collect and assess lessons learned and best practices from all areas of the organization.

Standardize design, procurement, processes, assembly, and equipment throughout the supply chain. Limit the use of custom components: doing so reduces overall cycle time, simplifies training and tasks, reduces repeated mistakes, improves the opportunity for bulk discounts, and improves the opportunity for automation and operation standardization. Also limit exotic or unique components, as they drive higher prices due to low volumes and less supplier competition. Unique components also result in increased opportunities for supply chain disruptions.

Simplify the design by reducing the number of parts. Parts reduction is one of the best ways to reduce the cost of fabricating and assembling a product and increase quality and reliability. This activity results in fewer opportunities for defective parts and fewer assembly errors. Develop an approved or preferred parts lists or a standardized bill of materials (BOM), and directly link it back to the computer-aided design (CAD) system. When possible, use one-piece structures from injection molding, extrusions, castings, and powder metals or similar fabrication techniques instead of bolting or gluing multi-part assemblies. Establish part families of similar parts based on proven materials, architecture, and technologies that are scaled for size and functionality. Use parts that can perform more than one function: for example, a cover or base plate that also serves as a heat sink. Incorporate guiding, aligning, or self-fixturing features into housing and structures.

Design for lean processes. Lean supply, fabrication, and assembly processes are essential design considerations. Anything that does not add value to a product is waste and should be reduced or eliminated. Lean processes are more likely to be performed quickly and correctly, resulting in reduced throughput time, which equates to faster time to market and lower costs. Designs that are easy to assemble manually are more easily automated and assembly that is automated is more uniform, more reliable, and of higher quality.

8.6 Chapter 6: Design for Sustainability

Design for Sustainability (DfS), also known as Design for Life Cycle Management, is ultimately about designing to minimize obsolescence issues and ensuring that electronic products are testable and repair able. The decisions made during the design and development process greatly influence the ultimate life-cycle results and costs experienced by end-users. Design for Sustainability is a comprehensive design strategy that includes:

- Design for Reliability (DfR)
- Design for Manufacturability (DfM)
- Design for Repairability and Testability (DfT)
- Design for Excellence = DfR + DfM + DfS = optimized total ownership cost

A critical element of DfS is planning for the end-of-life and component obsolescence issues that are common for long-life, high-reliability products. Obsolescence management is a key driver that also mitigates the risk of counterfeit parts. The secondary market is frequently the supply chain of last recourse when a part goes obsolete or is under supply constraints. While it is possible to get high-quality, authentic parts in the secondary market, it is also possible to get nonconforming, reworked, or counterfeit components.

Companies should design products with a high level of integrity and assurance that there will be no loss of support during the design life of the equipment. Experience from operations provides a significant opportunity to feed lessons learned into the design process, to support the intention to design for integrity. This may include:

- Ensuring the design team understands and designs for all potential operating modes
- Identifying and assessing of all failure modes that lead to loss of containment
- Designing to incorporate monitoring and accessibility for inspection where possible
- Designing appropriate isolations and connections in place to simplify any intervention that may be required
- Recognizing potential operating philosophy in design reliability and maintainability (RAM) analysis

- Ensuring operations engineers attend the design failure mode, effects and criticality analysis (FMECA)

A company's FMECA study should identify all failures that could result in the loss of integrity. Companies should ensure inspection, monitoring, and intervention features are designed into the system to minimize the likelihood or the consequence of such failures occurring. Evidence should be provided that appropriate actions have been undertaken. Also estimate the level of risk reduction that the actions may bring about. Design a durable product such that no failures occur during storage, transport, or installation or during its operational life as a result of material degradation or unexpected loads arising from the environment to which the equipment is exposed during the various life cycle stages. The term *load* here includes:

- Types of equipment loading, including thermal, electrical, and mechanical
- Combinations of equipment loading, e.g. thermal and mechanical
- Rates of loading including static, cyclic and shock loading

Ensuring product durability may take the form of analysis or physical testing or both. Identify a program of analysis and/or testing to demonstrate the durability of the product assembly, taking into consideration potential modes and causes of failure during each stage of its life. Also submit evidence of all analyses and tests undertaken on the components used in the assembly, and specify load-limit criteria.

8.7 Chapter 7: Root Cause Problem Solving, Failure Analysis, and Continual Improvement Techniques

Root cause analysis (RCA) encompasses a category of problem-solving methods that focus on identifying the ultimate underlying reason of why an undesired event occurred. *Root cause analysis* is a generic term for diligent, structured problem-solving. Over the years, various RCA techniques and management methods have been developed. RCA is based on the belief that problems are most effectively solved by correcting or eliminating the root causes rather than merely addressing the obvious symptoms. The root cause is the trigger point in a causal chain of events. This trigger point may be natural or man-made, active or passive, initiating or permitting, obvious or hidden. Efforts to prevent or mitigate the trigger event are expected to prevent the outcome or at least reduce the potential for problem recurrence.

RCA is a comprehensive analysis method that identifies the chain of physical and human-related root cause(s). It differs from basic troubleshooting and problem-solving processes that typically seek solutions to specific, relatively simple problems. The undesired event may be a product durability failure, safety incident, customer complaint, quality defect, or human error. RCA focuses corrective and preventive action (CAPA) efforts on the actions that have the greatest potential impact. It is essential for guiding change management efforts in the most effective and cost-efficient direction.

The goal of RCA is preventing failures before they happen. Prevention can be accomplished via a thorough understanding and analysis of designs, materials, suppliers, and customers. When failures do occur, however, organizations also need methods that enable them to quickly learn from and respond to these events. Effective use of RCA enables the most economical use of resources to achieve these goals. RCA also drives the pursuit of actions most likely to result in success. All companies have constraints on people, money, and time. If an organization doesn't analyze, learn from, and prevent problems, it simply repeats them.

Continual improvement is essential in remaining competitive and in becoming and staying world-class. A company that consistently, reliably, and profitably provides the greatest value to customers succeeds. In today's competitive environment, only the survival of the fittest and the best is guaranteed. RCA is a critical factor in that success.

The authors hope that this book helps you to manufacture your products with better reliability and customer satisfaction.

Index

Design for Excellence in Electronics Manufacturing, First Edition.
Cheryl Tulkoff and Greg Caswell.
© 2021 John Wiley & Sons Ltd. Published 2021 by John Wiley & Sons Ltd.